Ab initio molecular orbital calculations for chemists

Ab initio molecular orbital calculations for chemists

Second edition

by

W. GRAHAM RICHARDS
Oxford University

and

DAVID L. COOPER
Harvard-Smithsonian Center for Astrophysics

CLARENDON PRESS · OXFORD · 1983

Oxford University Press, Walton Street, Oxford OX2 6DP
London Glasgow New York Toronto
Delhi Bombay Calcutta Madras Karachi
Kuala Lumpur Singapore Hong Kong Tokyo
Nairobi Dar es Salaam Cape Town
Melbourne Auckland
and associated companies in
Beirut Berlin Ibadan Mexico City Nicosia

Published in the United States by
Oxford University Press, New York

British Library Cataloguing in Publication Data

Richards, W. G.
Ab initio molecular orbital calculations for chemistry.—2nd ed.
1. Molecules 2. Wave functions
I. Title II. Cooper, David L.
541.2'8 QC174.2

ISBN 0-19-855369-2

Library of Congress Cataloging in Publication Data

Richards, W. G. (William Graham)
Ab initio molecular orbital calculations for chemists.

Includes bibliographical references and index.
1. Molecular orbitals. 2. Quantum chemistry.
I. Cooper, D. L. II. Title.
QD62.R37 1982 541.2'8 82-12426
ISBN 0-19-855369-2 (pbk.)

Printed in Hong Kong

PREFACE TO THE SECOND EDITION

Since the first edition of this book was written in the late 1960s, there have been both qualitative and quantitative changes in the subject. Our suggestion in the introduction that an *ab initio* molecular wavefunction computer program might come to be used as a black box produced near apoplexy in some reviewers; yet that is exactly what has happened. We also stand by the analogy drawn in the first edition: a molecular wavefunction package can be used by an experimental chemist just as he or she may use a spectrograph. The analogy is fairly precise. The most successful applications will be made by those who understand the workings of the black box even if they cannot produce their own versions. Care has to be taken in preparing the input data and in interpreting the output, since computers may be the most treacherous of instruments.

Ab initio molecular orbital calculations are now a thriving industry as evidenced by the growth of our compilation of calculations. The weight of the subject has shifted towards larger molecules and the use of sophisticated, freely available, universal programs.

The new entrant to this field, who is likely to be an experimentalist rather than a full-time theoretician, needs a brief introduction, possibly something nearer to a cookery book than a standard monograph. The first edition seems to have fulfilled this role but is now so out of date that it presents a distorted view of the contemporary scene. Currently, gaussian basis sets are more widely used than exponentials; open-shell calculations are not so rare; the correlation energy problem is receiving serious attention. These topics are reflected in the changes which we have made.

Some necessary background is given, but as little mathematics as possible is included and proofs are omitted since they may be found in a number of texts on molecular quantum mechanics. It is assumed that the reader will have followed a course in elementary quantum chemistry, including the applications of group theory. Much use is made of worked examples, employing BH, CO, N_2, H_2O, NH_3, H_2CO, and MF_6.

The growth in work of this type shows no sign of slackening and is being encouraged by developments in computing which were undreamed of when the first edition was published. Our hope is that this edition will enable chemists and molecular physicists to reap full advantage of the new computer power.

We should like to thank John Horsley who co-authored the first edition for his encouragement to produce this new version.

Oxford and Harvard W.G.R.
January 1982 D.L.C.

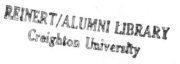

CONTENTS

1

INTRODUCTORY SUMMARY OF QUANTUM MECHANICS

This short chapter makes no attempt at completeness. It merely tries to indicate the relevant parts of quantum mechanics required for molecular calculations and to introduce notation.

1.1 The Schrodinger equation

All molecular orbital calculations are approximate solutions of the Schrödinger equation. This famous equation can be rationalized, although not derived, as an extension of the simple de Broglie equation to conditions where potential energy intervenes. The de Broglie relation

$$p = \frac{h}{\lambda}$$

tells us that radiation of wavelength λ is associated with a photon of momentum p, with h being Planck's constant. Planck's constant was introduced by him in explaining the experimental black-body radiation curves by means of the equation

$$E = h\nu,$$

which first postulated the quantization of energy. The frequency of radiation ν is related to the wavelength by

$$c = \nu\lambda,$$

c being the velocity of light.

Schrödinger's equation can be thought of as a combination of the de Broglie relation with the classical differential equation describing the profile of a simple harmonic three-dimensional standing wave:

$$-\frac{\hbar^2}{2m}\nabla^2\psi + V\psi = E\psi.$$

Here ψ gives the profile of the wave associated with a particle of mass m moving in a field of potential V and having energy E, with $\hbar = h/2\pi$. $\nabla^2\psi$ (read as 'del squared psi') is the laplacian operator;

$$\left(\nabla^2 = \frac{\partial^2}{\partial x^2} + \frac{\partial^2}{\partial y^2} + \frac{\partial^2}{\partial z^2}\right).$$

This wave equation, the time-independent Schrödinger equation, is frequently written in the almost cryptic shorthand form

$$\mathrm{H}\psi = E\psi,$$

in which the hamiltonian operator H is representative of the sum of the kinetic and potential energy of the system. (In this book, 'the system' will generally be an isolated molecule or more precisely a set of electrons moving in the field of the nuclei.) The equation has solutions ψ_i which are *eigenfunctions* of the operator H, alternatively called 'wavefunctions', and *eigenvalues* E_i, which are allowed quantized total energies of the system.

By analogy with light waves, the probability per unit volume of finding the particle at a given point is proportional to the square of the amplitude of the wave at that point. Thus in quantum mechanics $\psi^2 \mathrm{d}v$ is a measure of the probability or electron density for a one-electron wave in volume $\mathrm{d}v$. Since the function ψ may be complex a more correct measure of probability is $\psi^* \psi \mathrm{d}v$, ψ^* being the complex conjugate of ψ.

1.2 Wavefunctions

A wavefunction ψ is just a mathematical function like any other. However, for the remarks made above about the connection between ψ and probability to be meaningful, ψ must be single-valued, finite, and continuous at any point in space, and moreover be a quadratically integrable function which goes to zero at infinity in the case of a bound state. Solutions of the Schrödinger equation that correspond to different eigenvalues are orthonormal so that there is overall unit probability of the particle being somewhere in any particular state, i.e.

$$\int \psi_i^* \psi_i \mathrm{d}v = 1, \text{ but} \int \psi_i^* \psi_j \mathrm{d}v = 0.$$

In quantum mechanics there can be assigned to any closed physical system a wavefunction which can be operated on by an appropriate operator to give a value of any physical observable of the system as an eigenvalue of the operator equation (provided the quantity is a constant of motion). If more than one eigenstate corresponds to one eigenvalue then any new state formed by superposition (i.e. linear combination) of the original states will also be an eigenstate with the same eigenvalues; the state of the system is said to be degenerate. In the case of the hamiltonian operator, such states have the same energy. Often degeneracy may be removed by external applied fields which introduce new terms into the hamiltonian.

In the case of the electron bound in the hydrogen atom, the function that describes the properties of the system may be represented in polar coordinates as

$$\psi_{nlm}(r, \theta, \phi) = R_{nl}(r) Y_{lm}(\theta, \phi),$$

where $R_{nl}(r)$ governs the radial part of the wavefunction and nlm are the familiar integral quantum numbers. The spherical harmonics $Y_{lm}(\theta, \phi)$ govern the angular variation of the wavefunction.

The well-known drawings of hydrogen-atom wavefunctions or orbitals, e.g.

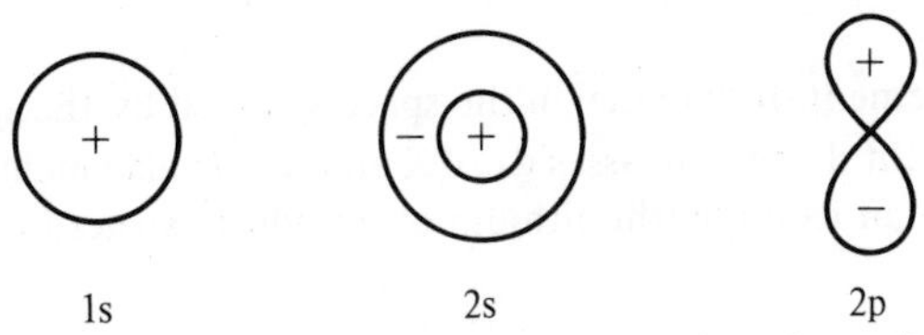

are merely two-dimensional representations of the three-dimensional wavefunctions. The signs refer to the radial part, i.e.

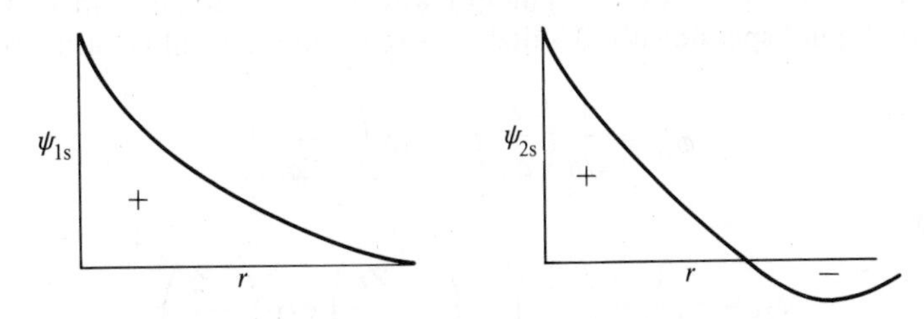

The three-dimensional representations frequently drawn in elementary chemistry books are drawings of $\psi^*\psi$ which is a measure of probability or electron density for a one-electron wavefunction. Note that this quantity is always positive.

Molecular orbital theory treats molecules from the same point of view as that which we use for atoms. In a many-electron system we make an approximation known as the orbital approximation. We assume that each electron has its separate wavefunction or orbital; each of these is an eigenfunction of a one-electron hamiltonian operator.

An orbital is then just an ordinary mathematical function describing the properties of one electron in a molecule.

1.3 The Pauli Principle

A fourth quantum number, the spin quantum number $M_s = \pm\frac{1}{2}$, is necessary to account for the doublet structure of spectral lines for alkali metal atoms. Spin occurs naturally in the solution of Dirac's modification of the Schrödinger equation which includes relativistic changes in mass.

If we include spin in our one-electron wavefunctions then we have spin orbitals. The spatial part is multiplied by a spin function α or β; the total wavefunction is now a product of spin orbitals, one for each electron. The one-electron spin wavefunctions are orthonormal, i.e.

$$\int \alpha^* \alpha ds = \int \beta^* \beta ds = 1$$

$$\int \alpha^* \beta ds = \int \beta^* \alpha ds = 0,$$

where ds is an element of 'volume' in the space spanned by the spin coordinates. The volume element dv encompasses dx, dy, and dz; dτ also includes spin, ds.

Let us take as an example the lithium atom whose structure is $1s^2 2s$ and has the wavefunction

$$\psi = \phi_{1s}(1)\bar{\phi}_{1s}(2)\phi_{2s}(3)$$

where the (1), (2), and (3) refer to the three electrons. A function written without a bar is multiplied by α ('spin-up') whereas an overlined function corresponds to β spin ('spin-down'). A suitable 1s wavefunction will be of the form

$$\phi_{1s} = \frac{1}{\sqrt{\pi}}\left(\frac{Z}{a_0}\right)^{3/2} \exp\left(-\frac{Zr}{2a_0}\right);$$

similarly

$$\phi_{2s} = \frac{1}{\sqrt{(32\pi)}}\left(\frac{Z}{a_0}\right)^{3/2}\left(2-\frac{Zr}{a_0}\right)\exp\left(\frac{-Zr}{2a_0}\right),$$

where Z is the nuclear charge and a_0 is the Bohr radius.

Even having included spin, our wavefunction is not yet complete. The Pauli Principle applies to any wavefunction describing a collection of particles under exchange of any pair of particles: the wavefunction must keep the same sign for particles with integral spin (bosons) and must change sign for particles with half-integral spin (fermions). The electron is a fermion with spin $s = \frac{1}{2}$ and so the total wavefunction must be antisymmetric with respect to electron permutation. Hence in the above case, if we exchange electrons (1) and (2) then we would obtain a new function

$$\psi' = \phi_{1s}(2)\bar{\phi}_{1s}(1)\phi_{2s}(3)$$

so that the wavefunction is not antisymmetric with respect to exchange of this pair of electrons. The combination $(\psi - \psi')$ is satisfactory from this point of view; if we make the wavefunction antisymmetric with respect to *any* two electrons then we will need a total of six terms which may be conveniently written in the form of a determinant:

$$\Psi = \frac{1}{\sqrt{6}}\begin{vmatrix} \phi_{1s}(1) & \bar{\phi}_{1s}(1) & \phi_{2s}(1) \\ \phi_{1s}(2) & \bar{\phi}_{1s}(2) & \phi_{2s}(2) \\ \phi_{1s}(3) & \bar{\phi}_{1s}(3) & \phi_{2s}(3) \end{vmatrix}$$

This is called a Slater determinant; it is usual just to write down the diagonal and leave it as understood that the product has to be antisymmetrized and normalized:

$$\Psi = |\phi_{1s}\bar{\phi}_{1s}\phi_{2s}| \quad \text{or even } \Psi = |1s\overline{1s}2s|.$$

1.4 The expansion of orbitals

Our orbital ϕ, which in the present work will most frequently be a molecular orbital, is normally expanded in terms of a set of atomic orbitals (χ)

$$\phi = \sum_p c_p \chi_p.$$

because integrals involving atomic functions are rather easier than those involving molecular functions. The set of atomic orbitals is called a *basis set.* In practice, it is convenient if these are normalized:

$$\int \chi_i^* \chi_i \mathrm{d}v = 1$$

but they are not orthogonal

$$\int \chi_i^* \chi_j \mathrm{d}v = S_{ij} \quad \text{if } i \neq j.$$

The quantity S_{ij} is called an overlap integral, the atomic orbitals could be chosen to form an orthonormal set so that

$$\int \chi_i^* \chi_j \mathrm{d}v = \delta_{ij}.$$

The quantity δ_{ij} is called the kronecker delta and is equal to unity if $i = j$ and is equal to zero if $i \neq j$.

Our problem is to determine the coefficients in the linear combinations of atomic orbitals and thus to obtain a wavefunction. If we take the equation $\mathrm{H}\Psi = E\Psi$, multiply on the left by Ψ^*, and then integrate over all space and spin coordinates, we then obtain the formula

$$E = \frac{\int \Psi^* \mathrm{H} \Psi \mathrm{d}\tau}{\int \Psi^* \Psi \mathrm{d}\tau}.$$

In Dirac's bracket notation this is written as

$$E = \frac{\langle\Psi|\mathrm{H}|\Psi\rangle}{\langle\Psi|\Psi\rangle}\,.$$

If we chose an arbitrary set of expansion coefficients, how would we know how reasonable our wavefunction might be? The answer to this lies in the variation principle which tells us that the closer our wavefunction Ψ comes to the true wavefunction then the lower will be the energy obtained using the above equation (strictly this only applies to the lowest state of each symmetry). Our strategy will be to choose a set of coefficients which minimize the energy.

1.5 Matrix form of the Schrödinger equation

Consider a wavefunction Ψ expanded as a linear combination of basis functions χ_n. The expression given in the previous section for the energy can be written

$$E = \frac{\langle\Psi|\mathrm{H}|\Psi\rangle}{\langle\Psi|\Psi\rangle} = \frac{\sum_l\sum_m c_l^* c_m \langle l|\mathrm{H}|m\rangle}{\sum_l\sum_m c_l^* c_m \langle l|m\rangle}\,.$$

To find the values of the coefficients which lead to the minimum energy, we differentiate with respect to c_n^* and set the derivative $\partial E/\partial c_n^*$ equal to zero. This leads to

$$E\sum_m c_m \langle n|m\rangle = \sum_m c_m \langle n|\mathrm{H}|m\rangle$$

or

$$\sum_m (H_{nm} - ES_{nm})c_m = 0,$$

where we have written H_{nm} for

$$\int \chi_n^* \mathrm{H} \chi_m \,\mathrm{d}\tau$$

and S_{nm} for

$$\int \chi_n^* \chi_m \,\mathrm{d}\tau.$$

Proofs that the integrals H_{nm} and H_{mn}^* are equal (i.e. that the hamiltonian operator is hermitian) are given in several standard texts. These equations can be written in the compact matrix form

$$(\mathbf{H} - E\mathbf{S})\mathbf{c} = 0\,.$$

The terms *integral* and *matrix element* are generally taken to be synonymous throughout molecular quantum mechanics. The condition for a non-trivial solution is that the determinant of the matrix $(\mathbf{H} - E\mathbf{S})$ must be zero. This is the origin of the *secular determinant*

$$|\mathbf{H} - E\mathbf{S}| = 0.$$

Notice that if we choose combinations of basis functions which constitute an orthonormal set then $\mathbf{S}$ becomes the identity matrix (i.e. $S_{ij} = \delta_{ij}$).

In semi-empirical molecular orbital calculations the number of integrals to be computed is greatly reduced by, for example, restricting attention to valence electrons and by ignoring those two-electron integrals which depend on the overlapping of the charge densities on different atoms (neglect of diatomic differential overlap) or even by completely neglecting differential overlap arising from the charge densities of different basis functions. In Hückel theory, the precise form of the hamiltonian is not defined and semi-empirical estimates of integrals are used.

Ab initio literally means *from the beginning*: the hamiltonian is precisely defined (usually as the self-consistent field hamiltonian) and all matrix elements are calculated from first principles without use of empirical data.

Molecular orbital calculations always make use of the simple determinantal equation above. We shall see in the next section that symmetry considerations lead to some simplification of the secular equations.

1.6 Simplification of the secular equations

An object referred to a particular set of axes may be said to be symmetric if it is possible to transform the axes in such a way that the object and image are indistinguishable. Symmetry elements include mirror planes, inversion centres, and rotation axes.

The only possible non-degenerate wavefunctions of a molecule are either symmetric or antisymmetric with respect to symmetry operations since properties such as energy must be left unchanged by symmetry operations. In the case of degenerate functions, symmetry operations will lead to different amounts of mixing of the different wavefunctions.

Group theory is a powerful mathematical technique which is helpful in the handling of symmetry operations; it offers a systematic, almost mechanical, method of working out certain problems in rather less space and time than would otherwise be possible.

In the form of *character tables*, it records how the system is affected by symmetry operations – i.e. whether the sign of the wavefunction is changed and, in the degenerate case, how different functions are mixed. If we were considering, say, formaldehyde (H_2CO) and examined how the C=O bond was affected by the symmetry operations of the molecule then the information as to any changing of signs would constitute a *representation* of the symmetry group of the molecule. A particularly important representation of a group is the *totally symmetric representation* in which there are no sign changes resulting from any of the symmetry operations.

For a group with *n classes* of symmetry operations, all the possible representations of the group may be reduced to just *n* irreducible representations (using similarity transformations). The classification of electronic states in terms of spin and orbital angular momenta and symmetry properties uses the labels given to different irreducible representations. Thus, the $^1\Sigma^+$ ground state of HF has the symmetry properties of, or *transforms as*, the Σ^+ (or A_1) irreducible representation of the group $C_{\infty v}$ and has no resultant spin or orbital angular momentum. Similarly, a particular excited state of benzene has the label ^3B_u since this state transforms as the B_u irreducible representation of the group D_{6h}. The use of some of the standard results of group theory will be very useful in the calculations on molecules.

If some operator R commutes with the total hamiltonian H (i.e. $[H, R] = HR - RH = 0$) and if eigenfunctions of this operator are used to build matrix elements of the energy matrix, then non-zero elements occur only between functions that correspond to the same eigenvalue of R. This leads to considerable simplification of the secular equations. In molecular problems, the commuting operators are normally angular momentum operators or symmetry operators.

An integral $\langle\psi_1|R|\psi_2\rangle$ can only be non-zero if some part of it belongs to the totally symmetric representation. We are interested in the case where the operator is the hamiltonian, which transforms as the totally symmetric representation. There is a useful theorem in group theory that tells us that the integral

$$\int \psi_1^* H \psi_2 d\tau \equiv \langle\psi_1|H|\psi_2\rangle$$

will be equal to zero (that is, it will not contain the totally symmetric representation) unless ψ_1 and ψ_2 belong to the same irreducible representation. Thus, for example, there will be no non-zero elements of the energy matrix between Σ and Π states in a diatomic molecule. In other words, there are no off-diagonal matrix elements of the hamiltonian between Σ and Π states.

In this way, the set of secular equations may be simplified: large determinants are reduced to a series of blocks along the diagonal, e.g.

$$\begin{vmatrix} (H_{11}-E) & H_{12} & 0 & 0 & 0 \\ H_{21} & (H_{22}-E) & 0 & 0 & 0 \\ 0 & 0 & (H_{33}-E) & 0 & 0 \\ 0 & 0 & 0 & (H_{44}-E) & H_{45} \\ 0 & 0 & 0 & H_{54} & (H_{55}-E) \end{vmatrix} = 0$$

Each small block constitutes a separate secular equation for states of a particular symmetry and can be treated separately.

The matrix elements referred to here are between wavefunctions for complete states. Within the orbital approximation, these can be expressed as a sum of matrix elements between molecular-symmetry adapted spin orbitals which, in turn, can be expanded in terms of a linear combination of atomic orbitals of the same symmetry (or, indeed, any set of basis functions). We have seen that symmetry considerations reduce the secular equation to smaller problems. Furthermore, the symmetry properties of the basis functions (s, p_x, p_y, p_z, d_{z^2}, etc.) cause many of the atomic integrals to be identically zero.

2
MOLECULAR ORBITALS

In molecular orbital theory, the wavefunction for the molecule consists of an antisymmetrized product of orbitals; one orbital for each individual electron. This gives a Slater determinant. Furthermore, each of the one-electron orbitals is itself a complicated linear combination of atomic orbitals. We thus have a hierarchy of complexity.

The wavefunction of the molecule Ψ can be broken down as follows:

$$\Psi = \mathscr{A}\psi \qquad (\mathscr{A} \text{ is an antisymmetrizing operator}),$$

but

$$\psi = \phi_1\phi_2 \ldots \phi_n \qquad (\phi_i \text{ is a one-electron spin orbital, which contains a spin } \alpha \text{ or } \beta \text{ for each orbital}),$$

and the spatial part of

$$\phi_i = \sum_p c_{ip}\chi_p;$$

χ_p are atomic orbitals.

Thus although the use of the variation principle for the molecules will involve us in calculating the integrals $\langle\Psi|\mathrm{H}|\Psi\rangle$ and $\langle\Psi|\Psi\rangle$, these can be broken down into integrals involving the molecular orbitals, which in turn reduce, or, more honestly expand, to atomic integrals involving the atomic orbitals χ_p.

The detailed rules that simplify the calculation of such matrix elements will be treated fully in Chapter 4, but in this chapter some very simple systems will be treated in order to show how a molecular calculation will reduce to a problem of atomic integrals. Further, the examples will serve to introduce some more standard notation.

2.1 Atomic units

There is considerable advantage, even at the level of writing down equations, if one works in atomic units. These take as fundamental quantities

the mass of the electron	m_e	as the unit of **mass**,
the charge of the electron	e	as the unit of **charge**,
the Bohr radius	a_0	as the unit of **length** (called the bohr),

$e^2/4\pi\epsilon_0 a_0$ as the unit of **energy** (approximately 27.21 eV and called the hartree).

In these units Planck's constant $h = 2\pi$ and hence $\hbar = 1$ (the unit of action), and $8\pi^2 m/h^2 = 2$. Thus for the hydrogen atom the wave equation may be written

$$\left\{-\tfrac{1}{2}\nabla^2 - \frac{1}{r}\right\}\psi = E\psi.$$

Bound energies are negative numbers with zero on the energy scale having particles infinitely separated.

2.2 The hydrogen-molecule ion

The starting point for all molecular calculations is the Schrodinger equation

$$\mathrm{H}\Psi = E\Psi.$$

In this case if we label the nuclei A and B, so that r_A is the distance of the electron from A and r_B is the distance from B, we have for the wave equation

$$\left\{-\tfrac{1}{2}\nabla^2 - \frac{1}{r_A} - \frac{1}{r_B}\right\}\psi = E_{\mathrm{el}}\psi.$$

The three energy terms in the hamiltonian are

$-\tfrac{1}{2}\nabla^2$ the kinetic energy of the electron,

$-\dfrac{1}{r_A}$ the coulomb attraction between the nucleus A and the electron,

$-\dfrac{1}{r_B}$ the coulomb attraction between the nucleus B and the electron.

The energy E_{el} is thus the electronic energy. To this we need to add the nuclear repulsion, which in terms of atomic units is just $1/R$, i.e.

$$E_{\mathrm{total}} = E_{\mathrm{el}} + \frac{1}{R}.$$

The subscript on E_{el} is normally left out. E_{total} and E_{el} are negative numbers but $1/R$ is positive. The nuclei are invariably assumed to be fixed. We do separate calculations for each configuration of the nuclei. The calculated energy is the electronic part, to which is added the internuclear repulsion energy for the particular geometry chosen. (This separation of energies is a consequence of the Born–Oppenheimer approximation.)

2.3 Solution of the wave equation for H_2^+ by the LCAO method

There are several ways of setting about the solution of the wave equation for the hydrogen-molecule ion. For our purposes the method in which the molecular orbitals are chosen to be linear combinations of atomic orbitals is the most interesting. This is usually referred to as the LCAO method.

We represent our molecular orbital ϕ as a sum of the $1s$ atomic orbitals on the two nuclei. Since our problem concerns only a single electron there is no distinction between the wavefunction for the single electron ϕ and the molecular wavefunction ψ.

The two atomic orbitals will have equal coefficients by symmetry, so the orbital ϕ is

$$\phi = N(1s_A + 1s_B).$$

A and B refer to the two nuclei and N is a normalizing constant given by

$$N = \frac{1}{\sqrt{\{2(1+S)\}}},$$

where

$$S = \int 1s_A 1s_B \, dv.$$

The energy of the orbital and hence the electronic energy of the system will be given by the formula

$$E = \frac{\langle\psi|\mathsf{H}|\psi\rangle}{\langle\psi|\psi\rangle}.$$

Substituting the electronic hamiltonian for H_2^+ and the LCAO expression for ψ into this formula, we obtain

$$E = \frac{1}{2(1+S)} \int (1s_A + 1s_B) \left\{ -\tfrac{1}{2}\nabla^2 - \frac{1}{r_A} - \frac{1}{r_B} \right\} (1s_A + 1s_B) dv.$$

We have, by symmetry,

$$\int 1s_A(-\tfrac{1}{2}\nabla^2) 1s_A \, dv = \int 1s_B(-\tfrac{1}{2}\nabla^2) 1s_B dv$$

and

$$\int 1s_A \frac{1}{r_A} 1s_A dv = \int 1s_B \frac{1}{r_B} 1s_B dv.$$

The full expression for the orbital energy may therefore be simplified to

$$E = \frac{1}{(1+S)}\left[\int 1s_A\,(-\tfrac{1}{2}\nabla^2)\,1s_A \mathrm{d}v + \int 1s_A\,(-\tfrac{1}{2}\nabla^2)\,1s_B \mathrm{d}v - \right.$$

$$\left. - \int 1s_A \frac{1}{r_A}\,1s_A \mathrm{d}v - \int 1s_A \frac{1}{r_B}\,1s_A \mathrm{d}v - 2\int 1s_A \frac{1}{r_B}\,1s_B \mathrm{d}v\right].$$

For $1s_A$ and $1s_B$ we can use hydrogen-atom solutions

$$1s_A = \frac{1}{\sqrt{\pi}}\,\mathrm{e}^{-r_A} \text{ and } 1s_B = \frac{1}{\sqrt{\pi}}\,\mathrm{e}^{-r_B}.$$

This leaves us with a set of integrals to calculate. Some are trivial and some rather tedious, but all can be computed by standard methods in milliseconds on even very small computers. When the integrals have been computed one adds them up to find the energy.

2.4 The hydrogen molecule H_2

The hydrogen-molecule ion is very much a special case, since it only contains a single electron. Far more typical is the neutral molecule. The wave equation may be written in atomic units as

$$\left\{-\tfrac{1}{2}\nabla_1^2 - \tfrac{1}{2}\nabla_2^2 - \frac{1}{r_{1A}} - \frac{1}{r_{1B}} - \frac{1}{r_{2A}} - \frac{1}{r_{2B}} + \frac{1}{r_{12}}\right\}\Psi(1,2) = E\Psi(1,2),$$

where 1 and 2 refer to the two electrons and A and B to the two nuclei. Again $E_{\text{total}} = E_{\text{el}} + 1/R$, R being the internuclear distance.

The equation may be simplified by noting that apart from the $1/r_{12}$ term the hamiltonian is a sum of two H_2^+ hamiltonians, i.e.

$$\left\{\mathsf{H}_{(1)}^N + \mathsf{H}_{(2)}^N + \frac{1}{r_{12}}\right\}\Psi(1,2) = E\Psi(1,2).$$

If the electron repulsion was neglected and H just equalled $\mathsf{H}_{(1)}^N + \mathsf{H}_{(2)}^N$ we could replace $\Psi(1, 2)$ by a product of two one-electron functions ϕ_1 and ϕ_2. These one-electron functions or orbitals would be simply eigenfunctions of the equation

$$\mathsf{H}_{(1)}^N\phi_1 = \epsilon_1\phi_1,$$

where $\mathsf{H}_{(1)}^N$ is a hydrogen molecule ion hamiltonian.

This idea of building up molecular wave functions as products of one-electron solutions for H_2^+ corresponds exactly to the familiar notion in atoms of saying 'the structure of the beryllium atom is $1s^2 2s^2$' – the $1s$ and $2s$ being derived from hydrogen-atom solutions.

For H_2 then we can say that the m.o. configuration is $1\sigma_g^2$ or $1\sigma_g^\alpha 1\sigma_g^\beta$ – two electrons with opposite spins in the lowest orbital of the type obtained in solving the wave equation for H_2^+. The orbital is labelled $1\sigma_g$ for symmetry reasons. In the notation introduced in the last chapter we could write the wavefunction as a Slater determinant

$$\Psi = |1\sigma_g 1\bar{\sigma}_g|.$$

This is the shorthand form of

$$\frac{1}{\sqrt{2}}\begin{vmatrix} 1\sigma_g(1) & 1\bar{\sigma}_g(1) \\ 1\sigma_g(2) & 1\bar{\sigma}_g(2) \end{vmatrix} \equiv \frac{1}{\sqrt{2}}\,[1\sigma_g(1)1\bar{\sigma}_g(2) - 1\sigma_g(2)1\bar{\sigma}_g(1)].$$

The molecular orbital $1\sigma_g$ could be expressed in LCAO form as

$$\frac{1}{\sqrt{2}}(1s_A + 1s_B).$$

However we may add even more terms to this expansion of $\phi_{1\sigma_g}$ if we wish,

e.g. $$\phi_{1\sigma_g} = c_1 1s_A + c_2 1s_B + c_3 2s_A + c_4 2s_B + c_5 2p_{xA} + \ldots,$$

but whatever the length of the LCAO expression we can always express the energy (or any other expectation value) as a sum of one- or two-electron integrals over molecular orbitals ϕ_i. There are rules – Slater's rules (see Chapter 4) which enable this to be done quite simply even in complicated cases, but for H_2 the grand expression can be written out in full and it is illustrative to do so.

The molecular wavefunction Ψ is $|1\sigma_g 1\bar{\sigma}_g|$. Here $1\sigma_g$ is a linear combination of a.o.s, which are mathematical functions in space coordinates, generally polar coordinates r, θ, and ϕ, but here just functions of r, there being no angular variation, so that the function is symmetric about the molecular axis. This orbital becomes a spin orbital by multiplying by one of the two spin functions α or β, which are, as we have seen, orthonormal, i.e.

$$\int \alpha^*\alpha \mathrm{d}s = 1, \quad \int \beta^*\beta \mathrm{d}s = 1, \quad \int \alpha^*\beta \mathrm{d}s = \int \beta^*\alpha \mathrm{d}s = 0.$$

The energy of the ground state = $\langle\Psi|\mathrm{H}|\Psi\rangle$

$$= \tfrac{1}{2}\Big\langle 1\sigma_g(1)1\bar{\sigma}_g(2) -$$

$$- 1\sigma_g(2)1\bar{\sigma}_g(1)\Big|\mathrm{H}^N_{(1)} + \mathrm{H}^N_{(2)} + \frac{1}{r_{12}}\Big| 1\sigma_g(1)1\bar{\sigma}_g(2) - 1\sigma_g(2)1\bar{\sigma}_g(1)\Big\rangle.$$

We will now discuss the varius integrals individually.

The first term in the expansion of the above expression will be

$$\langle 1\sigma_g(1)1\bar{\sigma}_g(2)|\mathrm{H}^N_{(1)}|1\sigma_g(1)1\bar{\sigma}_g(2)\rangle.$$

If we separate electrons 1 and 2, this may be rewritten as

$$\int 1\sigma_g(1)\mathsf{H}^N_{(1)}1\sigma_g(1)d\tau_1 \int 1\bar{\sigma}_g(2)1\bar{\sigma}_g(2)d\tau_2,$$

since H^N_1 operates only on electron 1. Then if we further separate space and spin parts it becomes

$$\underbrace{\int 1\sigma_g(1)\mathsf{H}^N_{(1)}1\sigma_g(1)dv_1}_{\epsilon^N_{1\sigma g}} \times \underbrace{\int \alpha(1)\alpha(1)ds_1}_{1}$$

$$\times \underbrace{\int 1\sigma_g(2)1\sigma_g(2)dv_2}_{1} \times \underbrace{\int \beta(2)\beta(2)ds_2}_{1}.$$

Thus the whole integral is reduced to the singel term $\epsilon^N_{1\sigma_g}$ ($\epsilon^N_{1\sigma_g}$ is defined as $\int 1\sigma_g(1)\mathsf{H}^N_1 1\sigma_g(1)dv_1$). Similarly

$$\langle 1\sigma_g(1)1\bar{\sigma}_g(2)|\mathsf{H}^N_{(2)}|1\sigma_g(1)1\bar{\sigma}_g(2)\rangle = \epsilon^N_{1\sigma_g}.$$

There will be four such terms, so that when the sum of these integrals is multiplied by the factor $\frac{1}{2}$ we are left with $2\epsilon^N_{1\sigma_g}$. There then remain the terms involving the operator $1/r_{12}$. The first of these is

$$\left\langle 1\sigma_g(1)1\bar{\sigma}_g(2)\left|\frac{1}{r_{12}}\right|1\sigma_g(1)1\bar{\sigma}_g(2)\right\rangle.$$

Separating as before we get

$$\underbrace{\iint 1\sigma_g(1)1\sigma_g(2)\frac{1}{r_{12}}1\sigma_g(1)1\sigma_g(2)dv_1\,dv_2}_{J_{1\sigma_g 1\sigma_g}} \underbrace{\int \alpha(1)\alpha(1)ds_1}_{1} \underbrace{\int \beta(2)\beta(2)ds_2}_{1}.$$

$J_{1\sigma_g 1\sigma_g}$ is defined in this way, and if we write electron 1 on one side of the operator and electron 2 on the other it is clear that the integral represents the coulomb interaction between electron clouds due to the two electrons separated by a distance of r_{12}. Thus equivalent definitions are

$$J_{1\sigma_g 1\sigma_g} = \iint 1\sigma_g^2(1)\frac{1}{r_{12}}1\sigma_g^2(2)dv_1dv_2$$

$$= \iint 1\sigma_g^*(1)1\sigma_g(1)\frac{1}{r_{12}}1\sigma_g^*(2)1\sigma_g(2)dv_1dv_2$$

$$= \iint 1\sigma_g^*(1)1\sigma_g^*(2)\frac{1}{r_{12}}1\sigma_g(2)1\sigma_g(1)\mathrm{d}v_1\mathrm{d}v_2.$$

The cross term will vanish owing to spin orthogonality, e.g.

$$\left\langle 1\sigma_g(1)1\bar{\sigma}_g(2)\left|\frac{1}{r_{12}}\right|1\sigma_g(2)1\bar{\sigma}_g(1)\right\rangle = \iint 1\sigma_g(1)1\sigma_g(2)\frac{1}{r_{12}}1\sigma_g(2)1\sigma_g(1)\mathrm{d}v_1\mathrm{d}v_2$$

$$\times \underbrace{\int \alpha(1)\beta(1)\mathrm{d}s_1}_{0} \times \underbrace{\int \alpha(2)\beta(2)\mathrm{d}s_2}_{0}.$$

The two coulomb terms when multiplied by the factor $\frac{1}{2}$ give us a single J. The entire expression for the energy reduces to

$$E(\mathrm{H}_2 : {}^1\Sigma_g^+) = 2\epsilon_{1\sigma_g}^N + J_{1\sigma_g 1\sigma_g}$$

This type of procedure may be generally followed but where there are more electrons the number of terms is vast and a very large piece of paper is required. The final result is, however, very simple and perfectly understandable in words: 'the energy of H_2 is twice the sum of the energy we would have if there was only one electron in the molecule, plus the coulomb interaction between the two electrons' – a one-electron integral and a two-electron integral.

2.5 The triplet state of H_2

The ground state of $H_2(X\,{}^1\Sigma_g^+)$ has the configuration $1\sigma_g^2$. If we excite one of these electrons to the next unfilled orbital, $1\sigma_u$ and both electrons have the same spin, a configuration $1\sigma_g 1\sigma_u$, the resulting state is a triplet, ${}^3\Sigma_u^+$ being its spectroscopic designation.

The wavefunction $\Psi = |1\sigma_g 1\sigma_u|$, so that the energy will be given by $\langle 1\sigma_g 1\sigma_u | |\mathrm{H}| | 1\sigma_g 1\sigma_u |\rangle$, the denominator being unity. It is instructive to expand this as we did for the ground state since we encounter a further type of integral not found for the ground state. Since both electrons are of the same spin, say α, no terms will disappear owing to spin orthogonality. Thus we have

$$E = \tfrac{1}{2}\left\langle 1\sigma_g(1)1\sigma_u(2) - 1\sigma_g(2)1\sigma_u(1)\left|\mathrm{H}_{(1)}^N + \mathrm{H}_{(2)}^N + \frac{1}{r_{12}}\right|\right.$$

$$\left. 1\sigma_g(1)1\sigma_u(2) - 1\sigma_g(2)1\sigma_u(1)\right\rangle.$$

Most of the terms will behave just as the ground-state expansion except for the cross term

$$-\left\langle 1\sigma_g(1)1\sigma_u(2)\left|\frac{1}{r_{12}}\right|1\sigma_g(2)1\sigma_u(1)\right\rangle$$

$$=-\underbrace{\iint 1\sigma_g(1)1\sigma_u(2)\frac{1}{r_{12}}1\sigma_g(2)1\sigma_u(1)\mathrm{d}v_1\mathrm{d}v_2}_{K_{1\sigma_g 1\sigma_g}}$$

$$\times\underbrace{\int \alpha(1)\alpha(1)\mathrm{d}s_1}_{1}\times\underbrace{\int \alpha(2)\alpha(2)\mathrm{d}s_2}_{1}.$$

Alternatively $K_{1\sigma_g 1\sigma_u}$ may be written as

$$\iint 1\sigma_g(1)1\sigma_u(1)\frac{1}{r_{12}}1\sigma_g(2)1\sigma_u(2)\mathrm{d}v_1\mathrm{d}v_2,$$

the important thing to notice being that in a K (exchange) integral electron 1 is in two different orbitals as is electron 2, so that the interaction is not just a simple electrostatic repulsion.

The complete energy expression for the $^3\Sigma_u^+$ state is then

$$E(H_2:{}^3\Sigma_u^+)=\epsilon_{1\sigma_g}^N+\epsilon_{1\sigma_u}^N+J_{1\sigma_g 1\sigma_u}-K_{1\sigma_g 1\sigma_u}.$$

The two energy expressions obtained for the two states of H_2 illustrate the quite general form of energy expressions for molecules. The total energy is the sum of the one-electron energies (the energy each electron would have were it the only electron in the molecule) plus a coulomb integral (the interaction between the charge clouds) for every pair of electrons, minus an exchange integral for every pair of electrons in the molecule which have the same spin. The first two types of term can be immediately understood from purely classical electrostatic ideas, but exchange is a purely quantum mechanical effect with no physical meaning, resulting from the Pauli Principle, which ensures that the wavefunction is antisymmetric. It is the integrals between different product terms in the sum of products which give rise to K integrals as is shown in the above case.

2.6 Expansion of the molecular integrals

We have seen how the energy of the H_2 ground state is a sum of the two molecular integrals $\epsilon_{1\sigma_g}^N$ and $J_{1\sigma_g 1\sigma_g}$. These may be expanded in terms of atomic integrals.

If

$$1\sigma_g = \frac{1}{\sqrt{\{2(1+S)\}}}(1s_A + 1s_B),$$

then

$$\epsilon^N_{1\sigma_g} = \frac{1}{2(1+S)}\int(1s_A + 1s_B)\left[-\tfrac{1}{2}\nabla^2 - \frac{1}{r_A} - \frac{1}{r_B}\right](1s_A + 1s_B)dv;$$

similarly $J_{1\sigma_g 1\sigma_g}$ reduces (or rather increases) to

$$\frac{1}{4(1+S)^2}\int(1s_A + 1s_B)^2_{(1)}\frac{1}{r_{12}}(1s_A + 1s_B)^2_{(2)}dv_1dv_2.$$

Clearly if $1\sigma_g$ was expanded as a longer sum of atomic integrals and if there were many more electrons, the number of atomic integrals required would grow very rapidly. Indeed for quite small molecules the number of atomic integrals required may be as many as several million. This is why it was not until computers that were fast and had large stores became available that *ab initio* calculations made very much impact. However the basic theory is simple and its execution becomes routine. In their simplest form the programs are not very difficult to write, although this is not often necessary, as specialists have made their programs available. These computer programs involve the very efficient use of standard techniques and careful shunting of numbers from one part of the computer to another. However, the programs are now so sophisticated that many man-years are required for their development.

2.7 General LCAO method for diatomic molecules

In H_2^+ we constructed the $1\sigma_g$ m.o. by combining the two 1s hydrogen a.o.s. Pictorially,

$1\sigma_g = 1s_A + 1s_B \equiv$ (+) + (+) → (+)

and

$1\sigma_u = 1s_A - 1s_B \equiv$ (+) − (+) → (+|−)

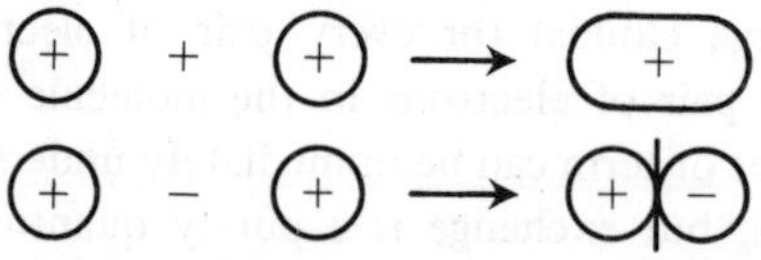

For a general homonuclear molecule we can represent the m.o.s. by the following well-known type of diagram given opposite.

The m.o.s are labelled σ or π depending on whether they are symmetrical about the internuclear axis or have a nodal plane passing through the nuclei and g or u depending on whether they remain unchanged or simply change sign when inverted at the centre of symmetry. This labelling is familiar, but the numbering $1\sigma_g$, $2\sigma_g$ $3\sigma_g$, etc. is a modern convention, replacing older versions which tended to indicate some atomic parentage of the m.o. This has been abandoned and the orbitals of a particular type are just numbered consecutively from the bottom of the diagram.

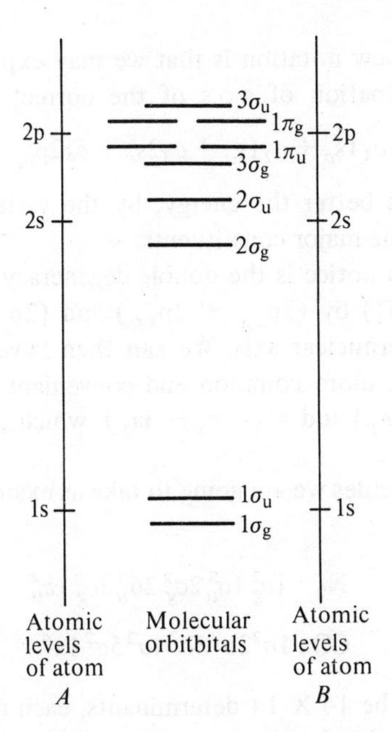

For heteronuclear molecules there is no longer a centre of symmetry, so there is no g or u, but there are still σ and π and the running number (see figure below).

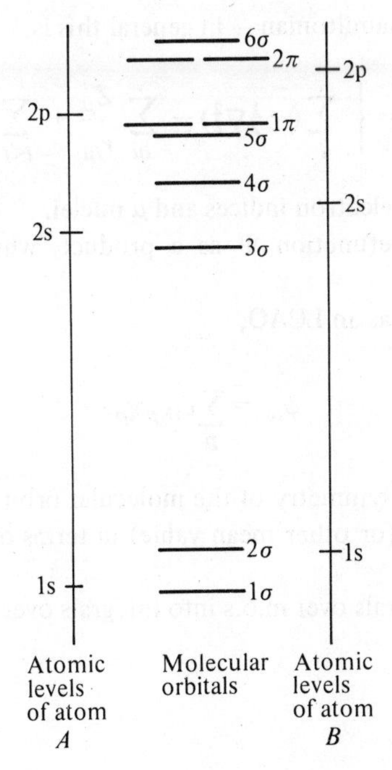

The reason for the new notation is that we may express for example 1σ as a very long linear combination of a.o.s of the correct symmetry, e.g. for CO

$$1\sigma = c_1 1s_c + c_2 1s_o + c_3 2s_c + c_4 2p_{\sigma_c} + \ldots$$

The more terms the better the energy, by the variation principle, but the less clear the origin of the major constituents.

One further point to notice is the double degeneracy of the π orbital formed in the simplest case (H_2^+) by $(2p_{x_A} + 2p_{x_B})$ and $(2p_{y_A} + 2p_{y_B})$, the z-axis being taken as the internuclear axis. We can then have π_x and π_y degenerate orbitals. It is, however, more common and convenient to work with the complex forms $\pi^+(= \pi_x + i\pi_y)$ and $\pi^-(= \pi_x - i\pi_y)$, which contain factors $e^{+i\phi}$ and $e^{-i\phi}$ respectively.

For two of the molecules we are going to take as examples the configurations of the ground states are

$$N_2 : 1\sigma_g^2 1\sigma_u^2 2\sigma_g^2 2\sigma_u^2 3\sigma_g^2 1\pi_u^4$$

$$CO : 1\sigma^2 2\sigma^2 3\sigma^2 4\sigma^2 5\sigma^2 1\pi^4$$

The wavefunctions will be 14 × 14 determinants, each term of which is a linear combination of atomic orbitals.

To formulate the energy expression for a molecule the procedure is as follows

(a) Write down the hamiltonian – in general this is

$$H = \left\{ \sum_i (-\tfrac{1}{2}\nabla_i^2) - \sum_{i\mu} \frac{Z_\mu}{r_{i\mu}} + \sum_{i<j} \frac{1}{r_{ij}} \right\}$$

where i and j are electron indices and μ nuclei.

(b) Express the wavefunction Ψ as a product, which has to be antisymmetrized.

(c) Express each $\phi_{i\lambda}$ as an LCAO,

$$\phi_{i\lambda} = \sum_p c_{i\lambda p} \chi_p.$$

Here λ labels the symmetry of the molecular orbital.

(d) Find the energy (or other mean value) in terms of integrals over molecular orbitals (ϕ_i).

(e) Expand the integrals over m.o.s into integrals over a.o.s (χ).

3
SELF-CONSISTENT FIELDS

We have seen that, if each m.o. is a linear combination

$$\phi_{i\lambda} = \sum_p c_{i\lambda p} \chi_p.$$

we can ultimately express all we need to calculate in terms of integrals involving the χ_p, multiplied, of course, by the appropriate sums and products of the LCAO coefficients $c_{i\lambda p}$. Indeed the only snag in the scheme outlined at the end of the previous chapter is, how do we know what the values of c are?

The answer to this is that we use the idea of self-consistent field orbitals.

3.1 The Hartree equations

The Hartree self-consistent field equations are based on the fact that if the wavefunction for the molecule is just a single product of orbitals, then the energy is the sum of the one-electron energies (kinetic energy and electron nuclear attractions) and coulomb interactions between the charge clouds of all pairs of electrons i and j.

$$E = \sum_i \epsilon_i^N + \sum_{i<j} \int \phi_i^2(1) \frac{1}{r_{12}} \phi_j^2(2) \mathrm{d}v_1 \mathrm{d}v_2.$$

The condition that this energy should be a minimum, an application of the variation theorem, together with the auxiliary conditions

$$\int \phi_i \phi_j \mathrm{d}v_i \mathrm{d}v_j = \delta_{ij}$$

gives us the Hartree equations for the 'best orbitals',

$$\left\{ \mathrm{H}^N + \sum_{j=1} \left(\int \phi_j^2(2) \frac{1}{r_{12}} \mathrm{d}v_2 \right) \right\} \phi_i(1) = \sum_j \epsilon_{ij} \phi_j(1),$$

each ϕ_i, being a linear combination of functions χ.

The hamiltonian contains terms involving the ϕ_i, which we wish to calculate. We thus have to use an iterative method of solution.

The Hartree equations are not now normally used since they are based on the idea of the wavefunction for the molecule being a single product of one-electron orbitals and not an antisymmetrized product.

3.2 The Hartree–Fock equations

These do consider the wavefunction to be an antisymmetrized product and enable one to reach the best Slater determinant solution iteratively.

As mentioned in the last chapter, if we include antisymmetry, then the energy is a sum of one-electron, coulomb, and exchange terms.

$$E = \sum_i \epsilon_i^N + \sum_{i<j} \int \phi_i^2(1) \frac{1}{r_{12}} \phi_j^2(2) \mathrm{d}v_1 \mathrm{d}v_2 -$$

$$-\sum_{i<j}{}' \int \phi_i(1)\phi_j(1) \frac{1}{r_{12}} \phi_i(2)\phi_j(2) \mathrm{d}v_1 \mathrm{d}v_2,$$

if all orbitals are assumed real. Using the more general Dirac notation, in the case of a closed shell molecule, this may be written as

$$E = 2\sum_k \epsilon_k^N + 2\sum_{kl} \left[\left\langle \phi_k\phi_l \left| \frac{1}{r} \right| \phi_k\phi_l \right\rangle - \tfrac{1}{2} \left\langle \phi_k\phi_l \left| \frac{1}{r} \right| \phi_l\phi_k \right\rangle \right],$$

where k and l now run over the doubly occupied m.o.s.

For closed-shell molecules, where all the electrons are paired with others of opposite spin,

$$E = \sum_i 2\epsilon_i^N + \sum_{i,j} (2J_{ij} - K_{ij}),$$

where here i and j label the orbitals; for example in the diatomic molecule BH, with $\Psi = |1\sigma^2 2\sigma^2 3\sigma^2|$,

$$E = 2\epsilon_{1\sigma}^N + 2\epsilon_{2\sigma}^N + 2\epsilon_{3\sigma}^N + J_{1\sigma 1\sigma} + 4J_{1\sigma 2\sigma} + 4J_{1\sigma 3\sigma} + J_{2\sigma 2\sigma} + 4J_{2\sigma 3\sigma} +$$

$$+ J_{3\sigma 3\sigma} - 2K_{1\sigma 2\sigma} - 2K_{1\sigma 3\sigma} - 2K_{2\sigma 3\sigma}.$$

(N.B. from the definition on p. 17 when $i = j$, $K_{ii} = J_{ii}$).

The Hartree–Fock equations are obtained by finding the condition for the energy to be a minimum $\delta E = 0$, and at the same time demanding that the molecular orbitals obtained shall be orthonormal,

$$\langle \phi_i | \phi_j \rangle = \delta_{ij}.$$

This use of the variation principle is strictly applicable only to the lowest state of a given symmetry.

The resulting equations are (for closed shells and real orbitals)

$$\left\{\mathsf{H}^N + \sum_{j=1}\int \phi_j^2(2)\frac{1}{r_{12}}\mathrm{d}v_2\right\}\phi_i(1) - \left\{\sum_{j=1}^{n}{}'\int \phi_j(2)\phi_i(2)\frac{1}{r_{12}}\mathrm{d}v_2\right\}\phi_j(1) = \epsilon_i^{\mathrm{SCF}}\phi_i(1)$$

or, in the shorthand form,

$$\left\{\mathsf{H}^N + \sum_j \mathsf{J}_j - \sum_j{}' \mathsf{K}_j\right\}\phi_i(1) = \epsilon_i^{\mathrm{SCF}}\phi_i(1).$$

Here we have used coulomb and exchange operators which are defined as

$$\mathsf{J}_j\phi_i(1) = \left(\int \phi_j^2(2)\frac{1}{r_{12}}\mathrm{d}v_2\right)\phi_i(1)$$

$$\mathsf{K}_j\phi_i(1) = \left(\int \phi_j(2)\phi_i(2)\frac{1}{r_{12}}\mathrm{d}v_2\right)\phi_j(1).$$

It should be noted that the J and K operators as opposed to J and K integrals have a single subscript rather than two.

In an even more compact form we may write

$$\mathsf{H}^{\mathrm{SCF}}\phi_i(1) = \epsilon_i^{\mathrm{SCF}}\phi_i(1).$$

Again since the hamiltonian contains the answer we are seeking, the set of equations, one for each ϕ_i, has to be solved iteratively.

For atoms where the same equations apply but we have the simplifying property of spherical symmetry, the Hartree–Fock equations can be solved numerically, giving an exact solution to the equations – not of course an exact solution to the problem, as there are assumptions inherent in the derivation of the equations which will be discussed later (Chapter 10). The exact solution of the equations is equivalent to taking our LCAO expansion to an infinite number of terms, and such a solution is called the Hartree–Fock limit.

Although some progress has been made towards solving molecular Hartree–Fock equations numerically, in general ϕ_i is represented as a linear combination and the equations are solved analytically. The longer the expansion used (the bigger the basis set) then, generally the closer the result will come to the Hartree–Fock limit.

3.3 The Roothaan equations

The Hartree–Fock equations in the LCAO approximation are normally called the Roothaan equations. A non-rigorous but readily comprehensible derivation is as follows.

We have, as above,

$$\mathsf{H}^{\mathrm{SCF}}\phi_i = \epsilon_i^{\mathrm{SCF}}\phi_i.$$

Now if

$$\phi_i = \sum_n c_{in}\chi_n$$

then

$$\mathsf{H}^{\mathrm{SCF}}\sum_n c_{in}\chi_n = \epsilon_i^{\mathrm{SCF}}\sum_n c_{in}\chi_n$$

and in the familiar manner we can multiply both sides of this equation by say χ_m and integrate over all space.

$$\sum_n c_{in}\int \chi_m \mathsf{H}^{\mathrm{SCF}}\chi_n \mathrm{d}v = \epsilon_i^{\mathrm{SCF}}\sum_n c_{in}\int \chi_m\chi_n \mathrm{d}v,$$

i.e.

$$\sum_n c_{in}(H_{mn}^{\mathrm{SCF}} - \epsilon_i^{\mathrm{SCF}} S_{mn}) = 0.$$

Such a set of equations is only soluble if

$$\det |H_{mn}^{\mathrm{SCF}} - \epsilon_i^{\mathrm{SCF}} S_{mn}| = 0.$$

This secular determinant looks very like that for the simple Hückel m.o. method where the matrix elements H_{mn} and S_{mn} are set equal to empirical parameters or even to zero.

If H_{mn}^{SCF} and S_{mn} could be calculated the secular determinant could be solved directly for the eigenvalues, the SCF orbital energies $\epsilon_i^{\mathrm{SCF}}$. However, both H_{mn} and S_{mn} demand a knowledge of the wavefunctions we are trying to find and yet again the solution has to be iterative, making the use of a computer mandatory.

3.4 The Roothaan-Hartree–Fock method

Because of its central importance in *ab initio* molecular orbital calculations we will now take a more detailed look at the working equations and their use in practice.

The *i*th orbital of symmetry λ may be degenerate so that we require an additional label α to identify the subspecies; for example, $i\lambda\alpha$ may represent $2\pi^+$. Each orbital $\phi_{i\lambda\alpha}$ is expanded as a linear combination of atomic basis functions $\chi_{p\lambda\alpha}$:

$$\phi_{i\lambda\alpha} = \sum_p c_{i\lambda p}\chi_{p\lambda\alpha}.$$

Notice that the expansion coefficients are independent of the subspecies – for example, coefficients for $2\pi^+$ and $2\pi^-$are identical.

The problem for a given basis set is to find values of the expansion coefficients, $c_{i\lambda p}$, which minimize the energy E. If we were to allow an infinitesimal variation of the coefficients, leading to a change δE in the energy, then this would be subject to the constraint that the orbitals must remain orthogonal. An important technique for finding the stationary points of a function of several variables subject to various constraint is Legrange's method of undetermined multipliers.

3.4.1 *Lagrange's method of undetermined multipliers*

For stationary points of $f(x, y)$

$$df = \frac{\partial f}{\partial x}dx + \lambda\frac{\partial f}{\partial y}dy = 0.$$

if, x and y are not actually independent but are related by $g(x, y) = 0$ then

$$dg = \frac{\partial g}{\partial x}dx + \frac{\partial g}{\partial y}dy = 0.$$

Hence, using a parameter λ,

$$d(f + \lambda g) = \left(\frac{\partial f}{\partial x} + \lambda\frac{\partial g}{\partial x}\right)dx + \left(\frac{\partial f}{\partial y} + \frac{\partial g}{\partial y}\right)dy = 0.$$

The lagrangian multiplier may be chosen such that

$$\frac{\partial f}{\partial x} + \lambda\frac{\partial g}{\partial x} = 0 \text{ and } \frac{\partial f}{\partial y} + \lambda\frac{\partial g}{\partial y} = 0.$$

These equations, together with the condition $g(x, y) = 0$, are sufficient to find the stationary points and the value(s) of the multiplier λ.

In the general case, there will be three classes of lagrangian multipliers, namely, those between closed-shell orbitals, those between open-shell orbitals, and those between closed- and open-shell orbitals. In the Roothaan–Hartree–Fock method, the last category of multipliers are removed by a suitable choice of separate operators, F_c and F_o, for closed and open shells. Note that the definitions of those operators include the expansion coefficients. The matrices for the other lagrangian multipliers can be diagonalized by separate unitary transformations on the closed shell and open shells. In matrix form, the equations to be solved take the form

$$\mathbf{F}_c\mathbf{c} = \boldsymbol{\epsilon}\mathbf{S}\mathbf{c}$$

$$\mathbf{F}_o\mathbf{c} = \boldsymbol{\epsilon}\mathbf{S}\mathbf{c}$$

where $\boldsymbol{\epsilon}$ is a transformed lagrangian multiplier matrix, $\mathbf{S}$ is an overlap matrix, and the elements of $\mathbf{c}$ are expansion coefficients. There are actually pseudo-eigenvalue equations since the matrices $\mathbf{F}_c$ and $\mathbf{F}_o$ are defined in terms of the solution $\mathbf{c}$. This is the origin of the need for a self-consistent field technqiue.

3.4.2 *Self-consistent fields*

The strategy frequently followed by SCF *ab initio* wavefunction programs is first to calculate all the various atomic integrals for a given basis set and geometry and to store these on some device such as a disk or tape. For small problems it may be possible to store the integrals in some compacted form in the core of the computer. Once all the integrals have been evaluated, and possibly reordered into a more convenient form, the SCF procedure is initiated.

An SCF procedure is represented schematically in the figure below. First of all, we must guess a set of expansion coefficients or trial molecular orbitals. These may indeed be any set of vectors which can be orthogonalised, such as the rows of the identity matrix. It is more usual to start with a more realistic guess so as to reduce the number of SCF iterations. This might come from chemical intuition as to the nature of different orbitals or from results of cal-

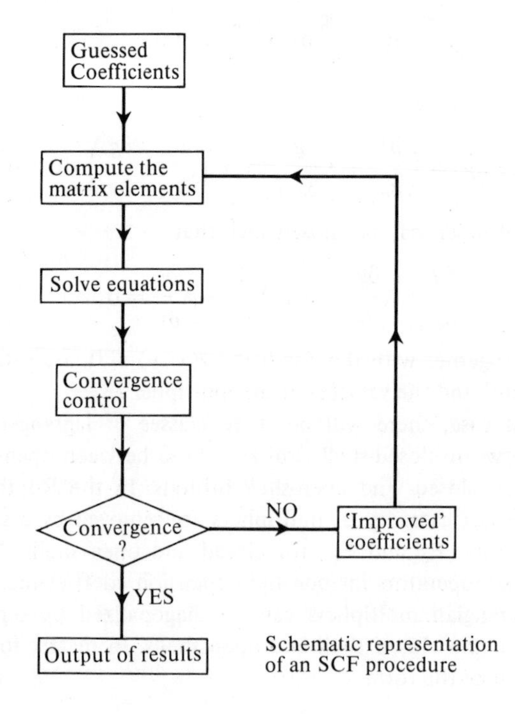

Schematic representation of an SCF procedure

culations on similar systems. Some programs can create their own set of trial molecular orbitals by, for example, diagonalizing the unscreened one-electron hamiltonian operator maxtrix.

Having formed some initial guess of the expansion coefficient we can now use the previously stored atomic integrals to construct the Roothaan–Hartree–Fock operators. Diagonalization and suitable reordering, usually in terms of energy, gives 'improved' expansion coefficients. This is the first cycle of an SCF procedure; the second and subsequent cycles proceed as above but use the coefficients generated by the previous iteration. An alternative procedure leading to equivalent results is to construct the operator in the basis of the trial molecular orbitals rather than in the basis of the atomic orbitals; in this case, the eigenvectors are 'improved' molecular orbitals expressed as linear combinations of the trial molecular orbitals rather than of the atomic basis functions.

The various available programs differ in how they test for convergence. However, whether they test, for example, the energies, or the coefficients or, perhaps, the mean square difference between successive density matrices (which are sums of products of coefficients formally defined later in Chapter 8) the principle is much the same. The SCF iterations are continued until the results from successive cycles are identical to within a given tolerance – say 10^{-6} in the coefficients.

An important question is: 'How do we know that the results from one iteration are better than those from the previous cycle and will convergence be uniform?' This is the reason for the box labelled 'convergence control' in the figure and the reason why the word 'improved' has been used in quotes. We shall discuss convergence in the next section.

3.5 Convergence

Left to its own devices, an SCF procedure for a closed-shell ground state may converge, albeit in a non-uniform manner, in a very small number of cycles. The first few cycles, particularly if the initial guess is rather arbitrary, tend to be somewhat erratic. Particularly for open-shell systems and for excited states, very slow convergence (perhaps with oscillatory behaviour) may be encountered. Worse still some systems exhibit fast divergence. Note that since the variation principle only applies to the lowest state of a given symmetry, excited states are frequently studied with some degree of 'configuration locking'; for example, 'improved' coefficients may be selected so as to produce maximum overlap with the trial values. Convergence may be rather slow for this sort of calculation.

Most programs provide some form of convergence control or some extrapolation procedure to deal with the above eventualities. Even in cases which will eventually converge of their own accord, some acceleration of the rate of convergence is desirable. The most obvious procedure is to look at the energies from successive iterations and to form some weighted average, or perhaps just the arithmetic mean, of the various sets of expansion coefficients. This generally leads to damping of any oscillatory behaviour and, since we can reject results leading to poorer (less negative) energies, it helps to avoid divergence. The extrapolation procedure employed by a particular program is often totally transparent to the user.

Some programs use rather more elaborate convergence control. For a Fock matrix constructed in the basis of the trial molecular orbitals, physically significant variations arise from the mixing of closed shells with open shells and with unoccupied ('virtual') shells and from the mixing of open shells with virtual shells. Mixing coefficients are determined, in part, by parameters known as damp factors and level shifters; some programs give the user some degree of control over these, and also some guidance as to suitable choices.

A further constraint that may aid convergence is symmetry. For example, some programs allow the user to specify whether a wavefunction is symmetric or antisymmetric to inversion through some centre.

In most cases, convergence is achieved with a minimum of effort in a fairly small (< 50) number of iterations. A few awkward cases, particularly those involving configuration locking, require some experimentation with initial guesses or with convergence parameters. A very small minority may be termed 'pathological'.

3.6 Optimization

A more detailed consideration of basis functions will be postponed for the time being. Optimal basis sets have been published for many atoms and molecules. Optimization of the orbital exponents for a given geometry is carried out automatically by computer programs using the variation principle. For a basis set of moderate size, the number of degrees of freedom is extremely large so that rather a large amount of computer time is required for optimization of all the exponents. This is the origin of the usefulness of published tables of basis sets.

Given a particular system, the other degree of system left to the user is the choice of geometry. Frequently, we may only be interested in one particular geometry. However, we may wish to know details of the potential energy as a function of some parameter such as a bond length or a bond angle. Some programs allow for *potential surface scans* by automatically repeating calculations at a series of geometries. Minima can be found by fitting, for example, quadratic expressions (parabolae) to energies at similar geometries.

Some programs can automatically carry out geometry optimization using the gradient of the Hartree–Fock energy. Analytical first derivatives of the Hartree–Fock energy are computed with respect to all the nuclear coordinates, and stationary prints on the potential surface found by minimization. There are alternative technqiues for this based on the method due to Fletcher and Powell or to Murtaugh and Sargent: the former technique does not require analytical derivatives of the energy while the latter does but is more reliable.

In outline, the cartesian forces and their derivatives are computed with respect to the internal coordinates. Using the current second-derivative (or hessian) matrix and the calculated first derivative vector, a projection is made to a new set of values for the variables. Each step in the optimization is composed of a linear search along the line connecting the current and previous points and a quadratic Newton–Raphson step. With the new geometry a new Hartree–Fock calculation is initiated but the wavefunction from the previous iteration suffices as an initial guess in the self-consistent field procedure.

When the optimization is completed the resulting geometry is normally very close indeed to experimentally determined bond lengths and angles.

3.7 An example – LiH

Let us take the simple example of LiH and outline the procedure. The atoms have the structures

$$\text{Li: } 1s^2 2s,$$
$$\text{H: } 1s$$

The m.o. diagram is then

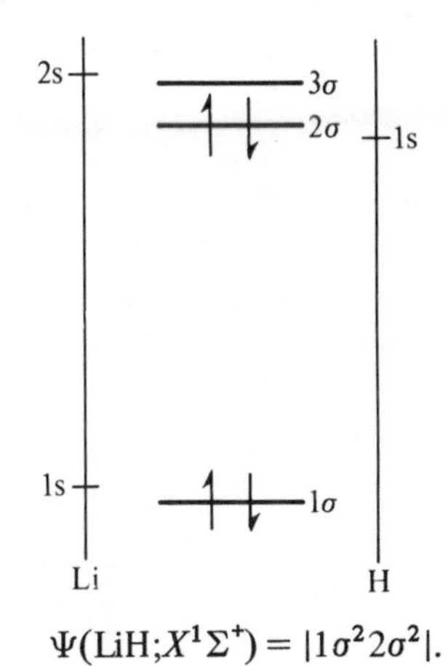

i.e. $$\Psi(\text{LiH};X^1\Sigma^+) = |1\sigma^2 2\sigma^2|.$$

Now

$$\phi_{1\sigma} = c_1 1s_{\text{Li}} + c_2 1s_{\text{H}} + c_3 2s_{\text{Li}},$$

where, for example,

$$1s = \frac{1}{\sqrt{\pi}} e^{-\alpha r} \text{ – an atomic orbital.}$$

In outline our procedure might be:

1. Calculate all the various integrals over atomic basis functions
2. Guess some values of the cs.
3. Build up $H_{i\sigma j\sigma}^{\text{SCF}}$ and $S_{i\sigma j\sigma}$
4. Solve the determinantal equation giving the possible values of ϵ_i^{SCF}
5. Substitute these in the secular equations giving new cs.
6. Go back to stage 3 and repeat until the values of ϵ_i^{SCF} or the cs converge to steady values within an arbitrary threshold ($\sim 10^{-6}$) and then take the values of the converged cs.

The result of this will be to provide SCF orbitals, both occupied and virtual (unoccupied): that is to say the coefficients in

$$\phi_{i\sigma} = c_{i1} 1s_{\text{Li}} + c_{i2} 1s_{\text{H}} + c_{i3} 2s_{\text{Li}} \text{ and } \epsilon_{i\sigma}^{\text{SCF}}.$$

This is all very straightforward, but lengthy, just the sort of problem at which computers are so powerful.

4
CALCULATION OF MATRIX ELEMENTS

In order to use the molecular orbitals found from a self-consistent field computation one must be able to calculate matrix elements such as the general energy expression

$$\frac{\langle\Psi|\mathrm{H}|\Psi\rangle}{\langle\Psi|\Psi\rangle}.$$

Here the denominator is unity if the wavefunction is normalized, and the capital letter Ψ represents a Slater determinant.

In Chapter 2 the evaluation of this expression for H_2 showed how many of the terms disappear due to spin or spatial orthogonality, and the result had a form with a clear rationality about it. To extend such expansions to cases with many more electrons would be fearsome were it not for the fact that these regularities persist and can be summarized in a set of rules which will be given and illustrated in this chapter. Before doing this, however, a few more words must be said about the integrals J and K.

4.1 Complex wavefunctions

For convenience in some of what has been said so far the molecular orbitals have been assumed to be real and the coulomb integral J_{ij} defined as

$$J_{ij} = \iint \phi_i^2(1)\frac{1}{r_{12}}\phi_j^2(2)\mathrm{d}v_1\mathrm{d}v_2.$$

This is the simplest way of seeing that it represents the coulomb interaction between the charge densities of electrons (1) and (2).

If the orbitals can be complex this is more properly written

$$\iint \phi_i^*(1)\phi_i(1)\frac{1}{r_{12}}\phi_j^*(2)\phi_j(2)\mathrm{d}v_1\mathrm{d}v_2,$$

which is the same thing as

$$\iint \phi_i^*(1)\phi_j^*(2)\frac{1}{r_{12}}\phi_i(1)\phi_j(2)\mathrm{d}v_{12};$$

electron (1) is still in orbital ϕ_i and (2) is in ϕ_j.

Another common shorthand is

$$\left\langle \phi_i\phi_j \left| \frac{1}{r_{12}} \right| \phi_i\phi_j \right\rangle,$$

where we understand that the complex parts are written on the left-hand side and the real parts on the right. Sometimes the operator $1/r_{12}$ is written as $g(i, j)$. A very useful shorthand form is the 'charge cloud notation'.

$$(ii|jj),$$

Where we keep electron (1) on one side and electron (2) on the other side of the operator which is not written in, and bear in mind that $\phi_i^2(1)$ is really $\phi_i^*(1)\phi_i(1)$ if this distinction is necessary.

This wealth of ways of expressing the same thing can cause confusion to those who are unfamiliar with the notation, so we now give all the common equivalent forms of writing $J_{2\sigma 1\sigma}$ and $K_{2\sigma 1\sigma}$

$$J_{2\sigma 1\sigma} = J_{1\sigma 2\sigma} = \iint \phi_{2\sigma}^*(1)\phi_{2\sigma}(1)\frac{1}{r_{12}}\phi_{1\sigma}^*(2)\phi_{1\sigma}(2)\mathrm{d}v_1\,\mathrm{d}v_2$$

$$= \langle\phi_{2\sigma}\phi_{1\sigma}|1/r_{12}|\phi_{2\sigma}\phi_{1\sigma}\rangle \text{ (or } \langle 2\sigma 1\sigma|1/r_{12}|2\sigma 1\sigma\rangle)$$

$$= (2\sigma 2\sigma|1\sigma 1\sigma)$$

and

$$K_{2\sigma 1\sigma} = K_{1\sigma 2\sigma} = \iint \phi_{2\sigma}^*(1)\phi_{1\sigma}(1)\frac{1}{r_{12}}\phi_{2\sigma}(2)\sigma_{1\sigma}^*(2)\ \mathrm{d}v_1\mathrm{d}v_2$$

$$= \langle\phi_{2\sigma}\phi_{1\sigma}|1/r_{12}|\phi_{1\sigma}\phi_{2\sigma}\rangle \text{ (or } \langle 2\sigma 1\sigma|1/r_{12}|1\sigma 2\sigma\rangle)$$

$$= (2\sigma 1\sigma|2\sigma 1\sigma).$$

In these examples $\phi_{2\sigma} = \phi_{2\sigma}^*$ as there are no complex parts, so that much of the difficulty is purely formal.

Care is, however, necessary when we do have complex parts. An m.o. $1\pi^+$ has an aximuthal portion $e^{+i\phi}$ and $1\pi^-$ had the factor $e^{-i\phi}$. This slight complication of course only applies to diatomic and linear polyatomic molecules where there is axial symmetry. Here we can distinguish two sorts of integrals involving π electroncs, usually labelled 2 or 0, i.e. containing the factors

$$e^{i\phi} \times e^{i\phi} = e^{2i\phi} \text{ or } e^{i\phi} \times e^{-i\phi} = e^{0\phi}.$$

Sometimes these cases are called + and −.

Thus

$$J_{1\pi 1\pi}^0 = \iint 1\pi^{+*}(1)1\pi^+(1)\frac{1}{r_{12}}1\pi^{+*}(2)1\pi^+(2)\mathrm{d}v_{12},$$

since

$$1\pi^{+*} = 1\pi^{-},$$

i.e.

$$J^{0}_{1\pi 1\pi} = (1\pi^{+}1\pi^{+}|1\pi^{+}1\pi^{+}).$$

If we use the latter shorthand we can see that the integral will be a J integral if the two spatial parts are identical for electron (1) and for electron (2). It will be a J^0 if the two exponential arguments are the same and J^2 if they are different. Likewise we can have K^0 and K^2 integrals

$$K^{0}_{2\pi 1\pi} = (2\pi^{+}1\pi^{+}|2\pi^{+}1\pi^{+})$$
$$K^{2}_{2\pi 1\pi} = (2\pi^{+}1\pi^{-}|2\pi^{+}1\pi^{-})$$

4.2 Rules for taking matrix elements

If we have a pair of wavefunctions written as Slater determinants

$$\Psi_1 = |1\sigma 1\bar{\sigma} 2\sigma 1\pi^{+}| \text{ and } \Psi_2 = |1\sigma 1\bar{\sigma} 1\pi^{+} 2\sigma|$$

how, if at all, do these differ? To answer this we must remember that the shorthand represents a determinant, and as we know from elementary mathematics, if we interchange two columns of a determinant we change the sign. Thus $\Psi_1 = -\Psi_2$.

It is convenient then when taking matrix elements between determinants firstly to bring the two determinants concerned into the maximum coincidence, remembering that each change of columns involves multiplying the wavefunction by (-1).

We have seen in Chapter 2 that the hamiltonian for a many-electron molecule consists of a sum of one-electron hamiltonians and repulsion terms between pairs of electrons. An n-body operator can at most change n spin-orbitals so that the matrix elements of an n-body operator can only be non-zero if the two states differ by no more than n spin-orbitals. A hamiltonian generally contains both one-electron and two-electron parts so that we need only consider states which differ from each other by at most two spin-orbitals.

There exists a set of general rules, called Slater's rules, for taking matrix elements between normalized Slater determinants, which we summarize in the next section.

4.2.1 *Slater's rules for matrix elements*

(i) *Identical determinants*

The matrix element of any one-electron operator, Ω_1, between identical determinants is given by

$$\langle \Psi_A | \Omega_1 | \Psi_A \rangle = \sum_i \langle i | \Omega_1 | i \rangle,$$

where the summation extends over all spin-orbitals.

The matrix element of two-electron operators, Ω_{12}, between identical determinants is given by

$$\langle \Psi_A | \Omega_{12} | \Psi_A \rangle = \sum_i \sum_{j \neq i} \{\langle ij | \Omega_{12} | ij \rangle - \langle ij | \Omega_{12} | ji \rangle\}.$$

Note that the integral vanishes when $i = j$.

(ii) *Determinants differing in one spin-orbital*

The matrix element of a one electron operation between determinants Ψ_A and Ψ_B which differ in one spin-orbital which is ϕ_i in the first case and ϕ_j in the second is given by

$$\langle \Psi_A | \Omega_1 | \Psi_B \rangle = \langle i | \Omega_1 | j \rangle.$$

For the case of a two-electron operator.

$$\langle \Psi_A | \Omega_{12} | \Psi_B \rangle = \sum_l \{\langle il | \Omega_{12} | jl \rangle - \langle il | \Omega_{12} | lj \rangle\}$$

with l running over all the other spin-orbitals occupied apart from i and j.

(iii) *Determinants differing in two spin-orbitals*

The matrix element of a one-electron operator between two determinants Ψ_A and Ψ_B which differ in two spin-orbitals which are $\phi_i \phi_j$ in the first case and $\phi_k \phi_l$ in the second is zero.

For a two-electron operator

$$\langle \Psi_A | \Omega_{12} | \Psi_B \rangle = \langle ij | \Omega_{12} | kl \rangle - \langle ij | \Omega_{12} | lk \rangle.$$

4.2.2 *Matrix elements of the hamiltonian operator*

(i) *Identical determinants*

Applying Slater's rules to matrix elements of the hamiltonian operator is very straightforward. The process is simplified by the fact that the hamiltonian is hermitian. The only possible source of confusion is that in applying Slater's rules the summations are over spin-orbitals, whereas it is sometimes convenient to consider and label doubly occupied orbitals. This may lead to the introduction of factors of 2.

Thus if we take a matrix element between two identical determinants but label the orbitals as follows

$$\Psi_A = |\phi_1^2\phi_2^2\phi_3^2 \ldots \phi_n^2|,$$

then

$$\langle\Psi_A|\mathrm{H}|\Psi_A\rangle = \sum_{k=1}^{2n} \epsilon_k^N + \sum_{k<l=1}^{2n} J_{kl} - \sum_{k<l=1}^{2n} K_{kl}.$$

The summations are over all spin-orbitals. The prime in this final summation indicates that exchange interactions are restricted to pairs of electrons of the same spin. Alternatively we could write the summations over doubly occupied molecular orbitals as

$$\langle\Psi_A|\mathrm{H}|\Psi_A\rangle = 2\sum_{i=1}^{n} \epsilon_i^N + 2\sum_{i,j=1}^{n} J_{ij} - \sum_{i,j=1}^{n} K_{ij}$$

or

$$\langle\Psi_A|\mathrm{H}|\Psi_A\rangle = 2\sum_{i=1}^{n} \epsilon_i^N + 4\sum_{\substack{i\leqslant j\\ i=1\\ j=1}}^{n} J_{ij} - 2\sum_{\substack{i\leqslant j\\ i=1\\ j=1}}^{n} K_{ij}.$$

Examples

(a) BH

The wavefunction for the ground state of BH will be

$$\Psi = |1\sigma^2 2\sigma^2 3\sigma^2|.$$

Therefore using the rule above

$$E = \sum_{i=1}^{3} 2\epsilon_{i\sigma}^N + \sum_{i,j=1}^{3} (2J-K)_{i\sigma j\sigma}$$

This time we have written the summations in terms of m.o.s. For practice it is probably worthwhile expanding this by writing down every single term, e.g. $J_{1\sigma 2\sigma}$, etc., in order to verify that the above shortened form does include the correct number of integrals. In the expression we have let both indices i and j range from 1 to 3 so that both $J_{1\sigma 2\sigma}$ and $J_{2\sigma 1\sigma}$ (which are identical) are included. Further it should be remembered that $K_{i\sigma i\sigma} = J_{i\sigma i\sigma}$.

(b) CO

The energy of the ground state of CO, with

$$\Psi = |1\sigma^2 2\sigma^2 3\sigma^2 4\sigma^2 5\sigma^2 1\pi^4|,$$

$$E(^1\Sigma^+) = \sum_{i=1}^{5} 2\epsilon_{i\sigma}^N + 4\epsilon_{1\pi}^N + \sum_{i,j=1}^{5} (2J-K)_{i\sigma j\sigma} +$$

$$+ \sum_{i=1}^{5} (8J-4K)_{i\sigma 1\pi} + 6J_{1\pi 1\pi}^0 - 2K_{1\pi 1\pi}^2.$$

The last two terms could cause confusion so let us write out in full the interactions between the π electrons

$$1\pi^+ \overline{1\pi}^+ 1\pi^- \overline{1\pi}^-$$

there will be six coulomb interactions:

$1\pi^+$ with $\overline{1\pi}^+$
$1\pi^+$ with $1\pi^-$
$1\pi^+$ with $\overline{1\pi}^-$
$\overline{1\pi}^+$ with $1\pi^-$
$\overline{1\pi}^+$ with $\overline{1\pi}^-$
$1\pi^-$ with $\overline{1\pi}^-$

or diagrammatically

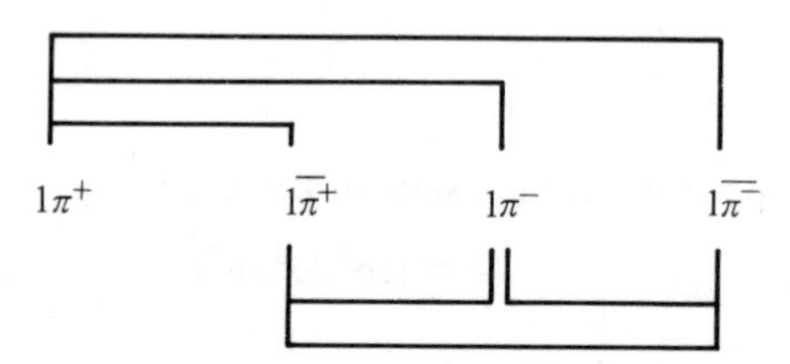

Each bracket represents an interaction.

For the K terms there are only two possible exchange interactions between electrons of the same spin,

$$1\pi^+ \text{ with } 1\pi^-, \text{ giving } K_{1\pi 1\pi}^2 = (1\pi^+ 1\pi^- | 1\pi^+ 1\pi^-)$$

and

$$\overline{1\pi}^+ \text{ with } \overline{1\pi}^-, \text{ giving } K_{1\pi 1\pi}^2 = (\overline{1\pi}^+ \overline{1\pi}^- | \overline{1\pi}^+ \overline{1\pi}^-).$$

This type of matrix element will always be necessary when calculating the energy of a particular state.

(ii) Ψ_A *and* Ψ_B *have one spin orbital different*

Our matrix element will be the integral

$$\int |\ldots \phi_k^\alpha \ldots |\mathrm{H}| \ldots \phi_l^\alpha \ldots | d\tau = \epsilon_{kl}^N + \sum_m (kl|mm) - \sum_m{}' (km|ml).$$

The summation is over spin-orbitals.

Here $\epsilon_{kl}^N = \int \phi_k^* \mathrm{H}^N \phi_l dv$. With these integrals particular care must be exercised in deciding which integrals are zero by reasons of spin orthogonality.

Example

Again using CO as an example, suppose we excite a $5\bar{\sigma}$ electron to a $6\bar{\sigma}$ spin orbital and take the matrix element between this function and the ground state, i.e.

$$\int |1\sigma^2 2\sigma^2 3\sigma^2 4\sigma^2 5\sigma 5\bar{\sigma} 1\pi^4 |\mathrm{H}| 1\sigma^2 2\sigma^2 3\sigma^2 4\sigma^2 5\sigma 6\bar{\sigma} 1\pi^4 | d\tau =$$

$$\epsilon_{5\sigma 6\sigma}^N + 2\sum_{i=1}^{4} (5\bar{\sigma}6\bar{\sigma}|i\sigma i\sigma) - \sum_{i=1}^{4} (5\bar{\sigma}i\bar{\sigma}|6\bar{\sigma}i\bar{\sigma}) + 4(1\pi 1\pi|5\bar{\sigma}6\bar{\sigma}) -$$

$$- 2(1\bar{\pi}6\bar{\sigma}|1\bar{\pi}5\bar{\sigma}) + (5\sigma 5\sigma|5\bar{\sigma}6\bar{\sigma}).$$

The summation is over m.o.s.

Such an integral will be needed when doing configuration–interaction calculations and will be returned to in Chapter 6 (p. 55).

(iii) Ψ_A *and* Ψ_B *differ by two spin orbitals*

In general our integral will be of the type

$$\int |\ldots \phi_k^\alpha \ldots \phi_p^\alpha \ldots |\mathrm{H}| \ldots \phi_l^\alpha \ldots \phi_q^\alpha \ldots | d\tau = (kl|pq) - (kq|pl).$$

We will only have both of these terms if all the four spin orbitals are of the same spin. Otherwise one or other integral will be zero for reasons of spin orthogonality.

Example

Again for CO, if both 5σ electrons are excited to the 6σ orbital, taking the matrix element with the ground state

$$\int |1\sigma^2 \ldots 4\sigma^2 5\sigma^2 1\pi^4 |\mathrm{H}| 1\sigma^2 \ldots 4\sigma^2 6\sigma^2 1\pi^4 | d\tau = (5\sigma 6\sigma|\bar{5}\sigma\bar{6}\sigma) - \underbrace{(5\sigma\bar{6}\sigma|\bar{5}\sigma 6\sigma)}_{\text{zero}}$$

the second term can be equated to zero for reasons of spin

$$\int \phi_{5\sigma}^{\alpha}(1)\phi_{6\sigma}^{\beta}(1)\frac{1}{r_{12}}\phi_{5\sigma}^{\beta}(2)\phi_{6\sigma}^{\alpha}(2)\mathrm{d}\tau =$$

$$\int \phi_{5\sigma}(1)\phi_{6\sigma}(1)\frac{1}{r_{12}}\phi_{5\sigma}(2)\phi_{6\sigma}(2)\mathrm{d}v \int \alpha(1)\beta(1)\mathrm{d}s_1 \int \beta(2)\alpha(2)\mathrm{d}s_2$$

$$= 0.$$

This type of integral is also of great importance in configuration interaction.

5
CLOSED-SHELL CALCULATIONS

We now have nearly enough information to consider performing a closed-shell computation and utilizing the results obtained. We must first, however, give some consideration to basis sets.

5.1 Basis sets of atomic orbitals

We are going to represent each molecular orbital $\phi_{i\lambda\alpha}$ as a linear combination

$$\phi_{i\lambda\alpha} = \sum_p c_{i\lambda p}\chi_{p\lambda\alpha}.$$

Since we require that the wavefunction should be single valued, finite, continuous, and quadratically integrable, it would seem a good idea to choose basis functions $\chi_{p\lambda\alpha}$ which also have these properties. The most obvious choice is a set of solutions of the Schrödinger equation for atoms since these would necessarily have the correct behaviour.

For a basis set of size m, there are $\frac{1}{2}m(m + 1)$ one-electron integrals and $\frac{1}{8}(m^4 + 2m^3 + 3m^2 + 2m)$ two-electron integrals. Even if m is quite small, it can be easily seen that the number of integrals is rather large; this leads to the need for large amounts of computer time. Furthermore, the time taken in the SCF cycling increases dramtically with the size of basis set and thus with the number of integrals. Indeed, if we were to proceed to electron-correlation calculations (see Chapter 10) we would require, as a first step, a transformation of the integrals which may depend on the fifth power of the number of basis functions.

It is clear that we should choose a basis set which is reasonably small. However, the basis set must not be so small as to affect accuracy adversely; it must be sufficiently flexible to describe the distortion resulting from molecule formation.

In the next few sections we shall examine the functional form of suitable orbitals to be used for a basis set.

5.2 Slater-type orbitals (STOs)

Solution of the Schrödinger equation for hydrogen-like atoms suggests the use of atomic orbitals of the form

$$\chi_{nlm} = r^{n-1}\mathrm{e}^{-\zeta r}Y_{lm}(\theta, \phi).$$

The radial variation consists of a power of r multiplied by an exponential function.

A 1s function thus depends on $e^{-\zeta r}$, a 2p function are $re^{-\zeta r}$, a 3d function on $r^2e^{-\zeta r}$, and so on. The angular part in θ and ϕ is a spherical harmonic $Y_{lm}(\theta, \phi)$. The exponent ζ ('zeta') is an adjustable parameter.

This type of function is usually normalized (by multiplying by a suitable constant) and is then called a Slater-type orbital (STO) or Slater-type function (STF). We shall use the name STO rather than STF since the term 'orbital' will serve to remind us that this function has the special properties we listed in the previous section.

We could thus construct a basis set for a many-electron atom by taking one or more STOs of the correct symmetry for each occupied orbital. If we take one exponential for each occupied orbital then we have what is frequently called a minimal basis set. In principle, there is no limit to the number of terms we take provided they are of the correct symmetry but we have seen above that very large basis sets necessitate considerable computational effort. Often two atomic orbitals with different exponents are taken for each occupied orbital; such a basis set is called a 'double zeta' (DZ) basis set. If we can take enough basis functions to reproduce the answer obtained by numerical solution then we are at the Hartree–Fock limit and, in general, by gradually augmenting a basis set, the energies converge on this result.

Basis sets of DZ (and larger) size have been published for many states of atoms and their ions. In each case, the orbital exponents ζ have been optimized so as to give the lowest energy. These atomic zeta values constitute a suitable starting point for the construction of basis sets for molecular calculations.

Ideally all the orbital exponents for a molecular system would be taken as variational parameters and the energy of the molecule would be computed for a series of values so as to find the minimum energy and thus the optimum zeta values. This sort of procedure is very time-consuming but has been carried out for many small molecules (such as all the diatomic hydrides of the first- and second-row atoms). More often, zeta values for molecules are taken from atomic calculations or, perhaps, from previous calculations on similar molecules. It is usually desirable to add to these functions, which for light atoms are of atomic s- and p-symmetry, a few d-symmetry orbitals whose exponents can be chosen by minimizing the energy with respect to these zeta values. This does not mean that the atomic d-orbitals are necessarily employed in bonding but simply that they are convenient variational functions. Particularly when chemical bonding is weak, the inclusion of polarisation functions (P) (d- and f-orbitals) in the basis set may lead to considerable energy improvement. Indeed, for a very weakly bound system, a DZ basis set may predict that the system is unbound while a DZ + P basis set may correctly predict a binding energy of, say, a few hundred wavenumbers. It is frequently also convenient specifically to augment the basis set for charged systems with basis functions for positively or negatively charged atoms: the addition of diffuse functions is particularly important for negative ions.

Particularly for small linear molecules, basis sets of DZ and DZ + P quality have been published and these can be easily modified for other molecules. The following table gives typical orbital exponents which have been used to compute molecular orbitals and expectation values for N_2 and CO.

Type	C	N	O
1sσ	5.303 60	6.212 92	7.165 12
1sσ	8.383 00	9.368 27	10.614 30
2sσ	1.269 60	1.467 86	1.601 11
2sσ	1.856 19	2.242 64	2.588 81
2pσ	1.287 09	1.528 53	1.651 45
2pσ	2.853 67	3.336 78	3.675 44
3dσ	1.895	1.935	2.103
2pπ	1.287 09	1.528 53	1.651 45
2pπ	2.835 67	3.336 78	3.675 44
3dπ	1.175	1.429	3.019

Source: R. K. Nesbet, *J. chem. Phys.* **40**, 3619 (1964).

STOs are used as basis functions for most accurate calculations on atoms and small, especially linear, molecules. The problem is that the many-centre two-electron integrals are rather difficult, require numerical integration techniques and are thus very time consuming. This is a severe problem for larger molecules and is the major reason for the introduction of a different type of basis function. Some progress has been made with the evaluation of multicentre integrals involving STOs through the Fourier transforms of the related B functions.

5.3 Gaussian-type orbitals (GTOs)

The cartesian gaussian orbitals of the form $x^l y^m z^n e^{-\alpha r^2}$, first suggested by Boys, have proved extremely useful in *ab initio* calculations of polyatomic molecules. The product of two GTOs is another GTO so that many-centre two-electron integrals reduce to much simpler forms. With gaussians all s-functions are taken to behave as $e^{-\alpha r^2}$ (cf. the STOs for 1s, 2s, and 2p functions in the previous section). Similarly, all p_z GTOs behave as $ze^{-\alpha r^2}$ and all d_{xy} GTOs behave as $xye^{-\alpha r^2}$.

The main disadvantage of the gaussian function is that it does not resemble very closely the form of real atomic orbital wavefunctions. In particular, the gaussian function lacks a cusp at the nucleus and hence the region near the nucleus is described rather poorly unless a large number of functions are used. The behaviour at large distances is also very different from that of the exact atomic orbital wavefunctions.

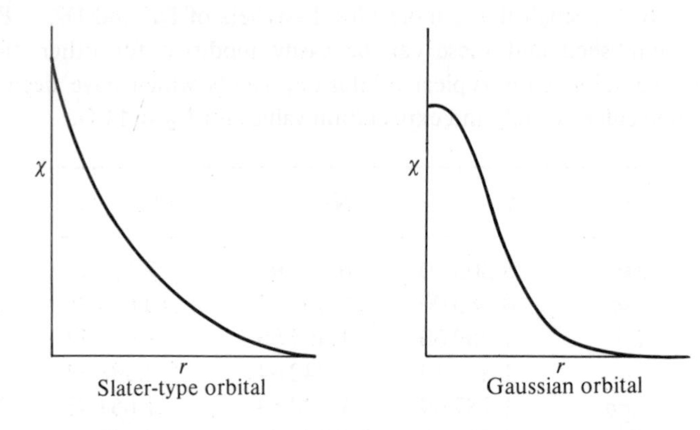

Slater-type orbital Gaussian orbital

A variational calculation of the energy of the hydrogen atom using a single gaussian function gives only 4/5 of the total energy, whereas a single exponential function gives the total energy exactly. This defect may be overcome by using a large number of gaussian functions with suitably chosen exponents in the basis set. However, this introduces difficulties in the solution of the Roothaan equations in the third part of the calculation. In particular, it becomes very difficult to get the iterative process to converge with a very large number of basis functions. Even if it is possible to obtain convergence, the time required to build up the matrix elements of the hamiltonian and diagonalize the resultant matrix increases enormously if a very large basis set is used.

The problem of the size of the basis set required when gaussian functions are used has been studied by Huzinaga. He found that the number of gaussian basis functions necessary is more than twice as great as the number of Slater-type exponential basis functions which gives an identical energy. Indeed a factor of about three is common. This somewhat depressing conclusion might lead one to believe that accurate calculations using gaussian functions would be extremely difficult. However, a method has been found to reduce the number of variables in the SCF calculation with very little loss of accuracy. Instead of allowing all the coefficients of the basis function expansion to vary freely, certain coefficients are fixed relative to one another, thus forming groups of gaussian functions, each known as a 'contracted gaussian', or CGTO.

The m.o. is then expressed as

$$\phi_{i\lambda\alpha} = \sum_p c_{i\lambda\alpha} \gamma_{p\lambda\alpha},$$

where $\gamma_{p\lambda\alpha}$ is a small contraction of gaussians of the same type on the same centre, e.g.

$$\gamma_1 = q_1\beta_1 + q_2\beta_2 + q_3\beta_3,$$

where we have used a single subscript to represent $p\lambda\alpha$.

In this way, a large basis set may be broken up into a much smaller number of groups. In the variational calculation of the molecular wavefunction only the coefficient of the contracted gaussian $c_{i\lambda p}$ is allowed to vary and not the relative proportions of the gaussians within each group (q_is). How much accuracy is lost as a result depends a great deal on the skill with which the initial basis of gaussians is contracted. The contraction process is largely a matter of using chemical intuition. For example, the 1s orbital of an atom in a molecule is unlikely to differ greatly from that in the isolated atom, so we can choose a group of gaussians having as coefficients those obtained for the 1s orbital in a variational calculation of the isolated atom. Similarly, in general it may be expected that only the long-range part of the 2s atomic orbital will be affected by the deformation caused by the bonding in the molecule. Therefore one can take another group of gaussians having as coefficients those of the 2s orbital in the isolated atom and then 'decomposing' this group into two groups, one containing the gaussian with the smallest exponent (longest range), and the other group containing all the other gaussians with the same relative proportions as in the isolated atom. These are just examples of the contraction process. In practice the extent to which the basis set is contracted will depend on the accuracy desired. Sets of variational calculations on atoms of the first and second rows and first transition series using gaussian functions has been published and the coefficients for the contracted gaussian groups are often taken from this work. The size of a basis set is often specified as, for example, a (7s3p/3s) primitive set contracted to [2s1p/1s].

As an example of a contracted gaussian basis set we take the basis set used by Clementi and Davis for an *ab initio* calculation on the ethane molecule. The exponents and coefficients for this basis set were taken from the set of calculations on the isolated atoms of the first-row elements carried out by Huzinaga. Using a basis set of 10 '1s' and 6 '2p' gaussian functions, Huzinaga minimized the energy of the carbon atom in the ^{3}P ground state with respect to variation of the exponents of the gaussian functions. The exponents and coefficients that he obtained are as overleaf.

Clementi and Davis grouped these functions together in the following way. To represent the region near the nucleus, which is poorly described by a single gaussian function, they grouped together the five functions $\beta_1 \rightarrow \beta_5$, all of which have rather large exponents. To represent the remainder of the 1s orbital they grouped together the two functions β_6 and β_7, which have the largest coefficients in the expansion of the 1s orbital in the atom. The 2s orbital, other than the region close to the nucleus, was represented by the three functions β_8, β_9, and β_{10}, which have large coefficients in the 2s orbital expansion. They did not contract these three functions into a single group but separated the function with the smallest exponent, β_{10}, from the pair of functions β_8 and β_9. In this way they partially allowed for the distortion of the 2s orbital upon bond formation. The six 2p functions were divided into two groups, a group formed from

1s

Function	Exponent	Coefficient
β_1	9470.52	0.000 45
β_2	1397.56	0.003 58
β_3	307.539	0.019 34
β_4	85.5419	0.077 36
β_5	26.9117	0.227 79
β_6	9.4090	0.426 95
β_7	3.500 02	0.357 91
β_8	1.068 03	0.048 77
β_9	0.400 166	−0.007 56
β_{10}	0.135 124	0.002 13

2s

Function	Exponent	Coefficient
β_1	9470.52	−0.000 10
β_2	1397.56	−0.000 76
β_3	307.539	−0.004 18
β_4	85.5419	−0.017 01
β_5	26.9117	−0.053 99
β_6	9.4090	−0.121 34
β_7	3.500 02	−0.175 54
β_8	1.068 03	0.085 02
β_9	0.400 166	0.606 89
β_{10}	0.135 124	0.438 09

2p

Function	Exponent	Coefficient
β_{11}	25.3655	0.008 75
β_{12}	5.776 36	0.054 79
β_{13}	1.787 30	0.182 63
β_{14}	0.657 71	0.358 71
β_{15}	0.248 05	0.432 76
β_{16}	0.091 064	0.203 47

the four functions $\beta_{11} \rightarrow \beta_{14}$ and a group consisting of the pair of functions with the smallest coefficients, again to make some allowance for the distortion of the 2p orbital upon formation of the molecule. Within each group of contracted gaussian functions the coefficients used were those obtained in the atomic calculation. The coefficients for the pair of functions β_8 and β_9 were taken from the 1s orbital expansion. The $2p_x$, $2p_y$, and $2p_z$ orbitals had identical exponents and coefficients.

We therefore have the following groups of contracted gaussian orbitals for the carbon atom:

Group 1: $0.000\,45\beta_1 + 0.003\,58\beta_2 + 0.019\,34\beta_3 + 0.077\,36\beta_4 + 0.226\,79\beta_5$
Group 2: $0.426\,95\beta_6 + 0.357\,91\beta_7$
Group 3: $0.085\,02\beta_8 + 0.606\,89\beta_9$
Group 4: β_{10}
Group 5: $0.358\,71\beta_{11} + 0.182\,63\beta_{12} + 0.054\,79\beta_{13} + 0.008\,75\beta_{14}$
Group 6: $0.432\,76\beta_{15} + 0.203\,47\beta_{16}$

The exponents and coefficients for the hydrogen atom were also taken from the atomic calculations of Huzinaga. They were taken from a four-term expansion in gaussian orbitals of a hydrogenic 1s orbital with exponent 1.0.

Exponent	Coefficient
0.123 317	0.509 07
0.453757	0.474 49
2.013 30	0.134 24
13.3615	0.019 06

To this expansion Clementi and Davis added a fifth gaussian function with exponent 0.07983. The four-term expansion formed one contracted gaussian group and the additional function was allowed to vary freely. More recently it has been found that the exponents of a group of gaussian functions representing a hydrogen orbital should be multiplied by a scale factor to take into account the contraction of this orbital in the molecular environment. A suitable scale factor for a single Slater-type orbital has been found to be 1.4 for a hydrogen atom bonded to a first-row atom. If the hydrogen orbital is represented by a group of gaussian functions this means that the exponent of every member of the group should be multiplied by a scale factor of 2 (see later).

We have seen that GTOs are computationally more convenient than STOs but have less appropriate functional form so that we require larger basis sets. Since we know that STOs are a much better description than GTOs, a sensible strategy to adopt would be to fit contractions or linear combination of gaussian functions to Slater orbitals. Least squares expansions of Slater orbitals of unit exponent into linear combinations of three, four or more gaussians have been published.

The important point is that the optimum coefficients for this linear combination are independent of the zeta value for the STO. The exponents of the GTOs which have been used to fit a STO with $\zeta = 1$ can be used to fit a similar STO by multiplying the exponent α by ζ^2 and then normalizing. In this way, contractions of gaussians may be stored in the program or on disk or tape and reused for all calculations. Some programs provide the user with basis sets of GTOs which have been fitted to STOs and can automatically construct minimal or extended basis sets from only information about atomic number.

The fact that GTOs can be used to model STOs has led to some progress in the calculation of two-electron integrals over Slater-type orbitals. The gaussian transform method of Shavitt and Karplus enables calculations to be carried out using a basis set of STOs while taking advantage of the special properties of GTOs. The method makes use of the integral transform

$$e^{-\alpha r} = \frac{\alpha}{2\pi}\int_0^\infty s^{-3/2}\, e^{-\alpha^2/4s}\, e^{-sr^2}\, ds\,.$$

Another set of contracted gaussians that has been used with considerable success is the 'gaussian-lobe' set proposed by Whitten. The gaussian-lobe functions avoid the computational difficulties involved in integrating over the angular part of the basis functions by expanding the functions that have angular dependence as linear combinations of simple gaussians without an angular part which are centred at different points in space. Thus a p-orbital is represented by a linear combination of two 'lobes' consisting of gaussians with radial dependence only, the centres of which are displaced an equal distance above and below the atomic nucleus, respectively.

5.4 Even-tempered and universal basis sets

The basis sets we have described in the last two sections are sufficient for most purposes. However, in very extensive calculations including correlation energy, the so-called basis set truncation error is probably the largest single source of error.

An even-tempered basis set consists of pure exponential or pure gaussian functions multiplied by a spherical harmonic. The orbital exponents ζ_k are defined by the geometric sequence

$$\zeta_k = \alpha\beta^k \qquad k = 1, 2, 3, \ldots, N$$

for each symmetry type. An even-tempered basis set thus contains only 1s, 2p, 3d, 4f, etc. functions. Values of α and β have been optimized for certain sizes of basis set (N) and may be obtained for other cases by simple recursion sequences. In principle, we could calculate energies for a series of basis sets of different size and then extrapolate to the basis set limit. A universal even-tempered basis set is one which is sufficiently large and flexible that it can be transferred from system to system with little loss of accuracy. Integrals need thus only be computed once and can then be reused for different atomic or molecular systems.

The principles involved in universal even-tempered basis sets are currently receiving much attention and progress looks promising.

5.5 An example of a closed-shell calculation

Having chosen an atomic orbital basis set, guessed coefficients, and used an *ab initio* SCF program, the resulting output will consist of the molecular-orbital coefficients (i.e. the final coefficients), their orbital energies $\epsilon_i^{\mathrm{SCF}}$, and possibly the molecular integrals.

The total electronic energy of the molecule is a sum of SCF orbital energies and one-electron energies for a closed-shell molecule.

This may be illustrated by taking $\mathrm{CO}(X^1\Sigma^+)$ again as an example.

$$\Psi = |1\sigma^2 2\sigma^2 3\sigma^2 4\sigma^2 5\sigma^2 1\pi^4|$$

Energy $= \langle\Psi|\mathsf{H}|\Psi\rangle$, where H is the electronic hamiltonian appropriate to the problem, the sum of kinetic and electrostatic terms.

As shown in Chapter 4 this integral is equal to

$$\sum_{i=1}^{5} 2\epsilon_{i\sigma}^{N} + 4\epsilon_{1\pi}^{N} + \sum_{ij=1}^{5}(2J-K)_{i\sigma i\sigma} + \sum_{i=1}^{5}(8J-4K)_{1\pi i\sigma} + 6J_{1\pi 1\pi}^{0} - 2K_{1\pi 1\pi}^{2}.$$

Now the SCF equation which has been solved is

$$\left\{\mathsf{H}^N + \sum_i \mathsf{J}_i - \sum_i \mathsf{K}_i\right\}\phi_j(1) = \epsilon_j^{\mathrm{SCF}}\phi_j(1). \tag{1}$$

This expression has been discussed above and is to be found in most books on quantum chemistry. It is, however, rather stark and possibly more readily understood by chemists by means of a specific example.

First it should be noted that σ- and π-orbitals are orthogonal for purely symmetry reasons, so that they are treated by separate equations, although terms which allow for interaction between σ- and π-electrons are included. Any π m.o. will contain only π-type a.o.s.

For the σ-electrons the SCF equation will be for CO

$$\left\{\mathsf{H}^N + \sum_{i=1}^{5}(2J_{i\sigma} - \mathsf{K}_{i\sigma}) + 4\mathsf{J}_{1\pi} - 2\mathsf{K}_{1\pi}\right\}\phi_\sigma = \epsilon_\sigma^{\mathrm{SCF}}\phi_\sigma. \tag{2}$$

The summation of $i = 1$ to 5 and the single 1π operator represent the fact that any electron will 'see' charge density due to those electrons which are in the molecule. Now we obtain an m.o. for every a.o. that has been taken in the basis, so that for a double zeta basis set we have about a dozen m.o.s. The hamiltonian is the same, however, for all m.o.s, both the filled ($i\sigma = 1$ to 5) and the unoccupied or virtual orbitals.

For the π-orbitals the wave equation in the SCF form is

$$\left\{\mathsf{H}^N + \sum_{i=1}^{5}(2J_{i\sigma} - \mathsf{K}_{i\sigma}) + 4J^0_{1\pi} - \mathsf{K}^2_{1\pi} - \mathsf{K}^0_{1\pi}\right\}\phi_\pi = \epsilon^{\mathrm{SCF}}_\pi \phi_\pi. \quad (3)$$

The two hamiltonians in (2) and (3) are often written for convenience in a block diagram to facilitate taking matrix elements of the SCF operator. The H^N will always be present and can be left out of the diagram as it is only in the J and K operators that confusion can arise.

The operators of eqns (2) and (3) can then be written as

Type	Occupied orbitals	
	$1\text{–}5\sigma^2$	$1\pi^4$
σ	$2\mathsf{J}-\mathsf{K}$	$4\mathsf{J}-2\mathsf{K}$
π	$2\mathsf{J}-\mathsf{K}$	$4\mathsf{J}^0-\mathsf{K}^2-\mathsf{K}^0$

The J and K are 'coulomb' and 'exchange' operators, although in some cases the integrals in $\langle\phi|\mathsf{H}^{\mathrm{SCF}}|\phi\rangle$ will not be strictly J or K integrals, i.e. not $(i\sigma i\sigma|j\sigma j\sigma)$ and $(i\sigma j\sigma|i\sigma j\sigma)$, but rather, for example $(i\sigma j\sigma|m\sigma m\sigma)$ and $(i\sigma m\sigma|j\sigma m\sigma)$ – sometimes called pseudo-J or pseudo-K integrals.

The form of the SCF operator is very important, since it enables one to express $\langle\phi|\mathsf{H}^{\mathrm{SCF}}|\phi\rangle = \epsilon^{\mathrm{SCF}}$ very simply as a sum of integrals.

Thus in our example

$$\epsilon^{\mathrm{SCF}}_{1\sigma} = \left\langle 1\sigma\middle|\mathsf{H}^N + \sum_{i=1}^{5}(2\mathsf{J}-\mathsf{K})_{i\sigma} + 4\mathsf{J}_{1\pi} - 2\mathsf{K}_{1\pi}\middle|1\sigma\right\rangle$$

$$= \epsilon^N_{i\sigma} + \sum_{i=1}^{5}(2J-K)_{1\sigma i\sigma} + 4J_{1\pi 1\sigma} - 2K_{1\pi 1\sigma},$$

or more generally,

$$\epsilon^{\mathrm{SCF}}_{i\sigma} = \epsilon^N_{i\sigma} + \sum_{j=1}^{5}(2J-K)_{i\sigma j\sigma} + 4J_{1\pi i\sigma} - 2K_{1\pi i\sigma}$$

and

$$\epsilon_{1\pi}^{\text{SCF}} = \epsilon_{1\pi}^{N} + \sum_{i=1}^{5} (2J-K)_{1\pi i\sigma} + 4J^{0}_{1\pi 1\pi} - K^{2}_{1\pi 1\pi} - K^{0}_{1\pi 1\pi}.$$

These orbital-energy expressions may be used to simplify the long energy expression

$$\langle \Psi | \mathsf{H} | \Psi \rangle$$

where Ψ is the Slater-determinant wavefunction and H the electrostatic hamiltonian.

If we do this, substituting $\epsilon_{i\sigma}^{\text{SCF}}$ and $\epsilon_{1\pi}^{\text{SCF}}$, we find that

$$\text{E} = \sum_{i=1}^{5} (\epsilon_{i\sigma}^{N} + \epsilon_{i\sigma}^{\text{SCF}}) + 2(\epsilon_{1\pi}^{N} + \epsilon_{1\pi}^{\text{SCF}}),$$

i.e. the energy of a closed-shell molecule is a simple sum of orbital and one-electron energies.

This result is quite general for all closed-shell molecules, both diatomic and polyatomic. If the electronic configuration is a set of doubly occupied orbitals labelled by the index i

$$E = \sum_{i} (\epsilon_{i}^{N} + \epsilon_{i}^{\text{SCF}}).$$

Note firstly that this is the electronic energy to which must be added nuclear replusion terms if we want, as we usually shall, the total energy of the molecule for a defined geometry. Secondly the superscript in $\epsilon_{i}^{\text{SCF}}$ is frequently not written and many books and papers use ϵ_i for the orbital energy.

The orbital energies, ϵ^{SCF}, and the one-electron energies, ϵ^{N}, which are the kinetic plus nuclear-attraction energies, are as we have seen negative numbers for bound electrons. Zero on the energy scale has all particles separated at infinite distances from each other. Calculated binding energies are thus negative numbers: the more negative the number the 'lower' the energy. More tightly bound electrons have more negative energies than the less strongly held ones. This simple point often causes confusion to novices who tend to think reasonably enough, of energy being 'low' at the 'bottom' of a potential energy curve or surface and increasing as one goes 'up'. The figure overleaf shows a computed curve for NH^{+} giving the quantum mechanical energies in atomic units and the spectroscopists' energy scale which runs in the opposite direction.

The virtue of the theoreticians' seemingly bizarre scale is that it has a common zero for all problems: one molecule may be compared directly with another in energy terms.

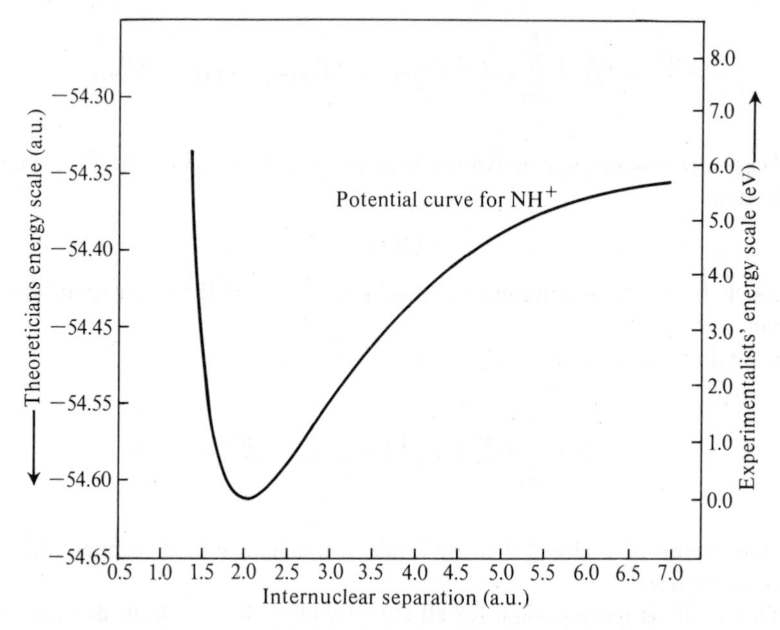

An interesting side-issue which is highlighted by the result that

$$E = \sum_i (\epsilon_i^N + \epsilon_i^{SCF})$$

is the use of the orbital energies, ϵ^{SCF}, in semi-empirical and empirical applications. Many semi-quantitative – and very useful – applications of molecular orbital theory depend solely on a consideration of the orbital energies, ignoring the one-electron, ϵ^N, terms. For example Walsh's rules enable predictions to be made about the shape of a molecule based on diagrams showing the variation of orbital energy with geometry. The figure opposite shows a typical diagram for AH_2 triatomic molecules.

Molecules with six electrons such as BeH_2 or BH_2^+ are predicted to be linear since the configuration $1a_1^2 2a_1^2 1b_2^2$ has a 'lower' energy at the right-hand side of the diagram whereas water with 10 electrons is bent. The latter result is explained by the strong variation in energy of the $3a_1$ molecular orbital which is more bonding ('lower') in a bent geometry.

The utility of these arguments is unquestionable, but their basis is less than secure. The result we have just shown above emphasizes that the ϵ^N terms should also be considered and not merely ϵ^{SCF}. In fact when done quantitatively the actual numerical values of ϵ^N are considerably greater than ϵ^{SCF} so that they tend to swamp the energy. What is more they have only slight variation with molecular geometry so that we are forced to conclude that Walsh's rules work due to a very fortuitous cancellation of errors.

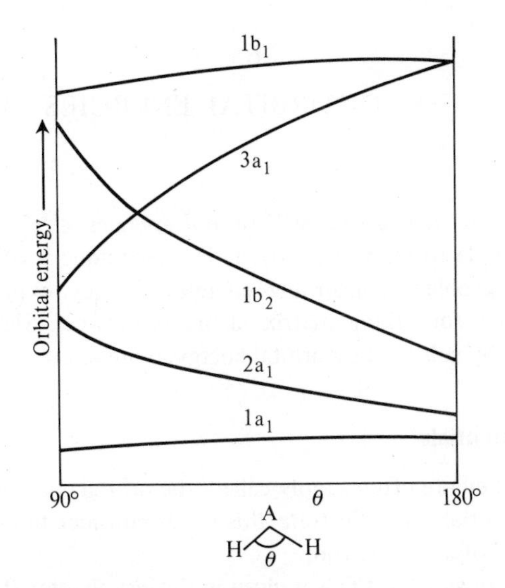

The technique by which the pleasingly simple result of energy being a sum of orbital and one-electron energies is obtained is important. Tremendous simplifications can be obtained if one does work in terms of ϵ^{SCF} rather than with long lists of integrals derived from the use of the electronic hamiltonian. Some applications of this idea are presented in the next chapter.

6
USES OF ORBITAL ENERGIES

Three applications of the use of SCF orbital energies, ϵ^{SCF}, to reduce energy expressions to simple forms will be given. The technique is of very general use for almost any problem concerning energies. Energy expressions obtained by using the rules for taking matrix elements between Slater determinants may be simplified by substituting orbital energy expressions.

6.1 Ionization potentials

The ϵ^{SCF}, or, as they are often simply called, the orbital energies ϵ, approximate to ionization potentials. To illustrate this let us consider the ionization of CO to CO^+ by removal of a 5σ electron.

The energy expression for CO was given in the last chapter. The wavefunction for the $^2\Sigma^+$ ground state of the ion will be

$$\Psi = |1\sigma^2 2\sigma^2 3\sigma^2 4\sigma^2 5\sigma 1\pi^4|.$$

The energy could be obtained by using the rule for taking matrix elements of the electronic hamiltonian using Slater's rules and comparing the energy expressions for the molecule and the ion. However, if we assume that the m.o.s do not change very much between molecule and ion then many of the terms will be the same in the two expressions. Indeed we see, simply by looking at the two wavefunctions, that with respect to the neutral molecule the ion will have the same energy expression, except that the following terms will not occur:

$$\epsilon_{5\sigma}^{N} + \sum_{i=1}^{4}(2J-K)_{5\sigma i\sigma} + J_{5\sigma 5\sigma} + 4J_{5\sigma 1\pi} - 2K_{5\sigma 1\pi}.$$

which is identical with $\epsilon_{5\sigma}^{\mathrm{SCF}}$.

This result is generally true for closed-shell molecules and is called Koopmans' theorem

$$\mathrm{I.p.}(i) \approx \epsilon_i^{\mathrm{SCF}}.$$

The table shows how some computed ϵ^{SCF} values agree with orbital energies found from photoelectron spectroscopy. The agreement is such that it is clear that *ab initio* m.o. calculations can be of considerable assistance in interpreting these data.

There are great dangers in the indiscriminate application of Koopmans'

Molecule	Orbital	Calc. (eV)	Photoelectron i.p. (vertical)
H_2	$1\sigma_g$	16.18	15.88
HF	1π	17.69	16.06
	3σ	20.90	20.00
CO	4σ	21.87	19.72
	5σ	15.09	14.01
	1π	17.40	16.91
N_2	$3\sigma_g$	17.28	16.96
	$2\sigma_u$	21.17	18.72
O_2	$3\sigma_g$	20.02	20.12
	$1\pi_u$	19.19	17.99
	$1\pi_g$	14.47	12.54

theorem. The assumption that all orbitals are unchanged when going from the molecule to the ion is clearly drastic, but further, implicit in the derivation of Koopman's theorem, is the notion that the difference in Hartree–Fock energies of ion and molecule is equal to the true difference of energies. This is by no means always the case, because of the inherent errors in the Hartree–Fock method. These errors, which can even result in predictions of the order of ionization potentials being in error, are discussed in Chapter 10. The classic case of error is N_2, where in the molecule the $1\pi_u$ electron has a smaller orbital energy than the $3\sigma_g$ electron, but none the less the ground state of the ion is $^2\Sigma_g^+$ as a consequence of correlation-energy differences in the ionic states.

6.2 Excitation energies

This same technique of taking matrix elements between determinantal wavefunctions (remembering as is discussed later that the wavefunctions may be linear combinations of determinants) and the use of ϵ^{SCF} to simplify the expressions can be used to find the excitation energies of other states of a molecule built using virtual orbitals.

Again taking CO as an example; by exciting a 5σ electron to a 6σ orbital we can obtain a $^3\Sigma^+$ state with the wavefunction

$$\Psi' = |1\sigma^2 2\sigma^2 3\sigma^2 4\sigma^2 5\sigma 6\sigma 1\pi^4|.$$

Now the energy of this state could again be found by taking the matrix element $\langle\Psi'|H|\Psi'\rangle$ and the excitation energy by subtracting $\langle\Psi|H|\Psi\rangle$, the energy of the molecule.

This is the safest manner in which to proceed, but it should soon become evident that this long-winded process is by-passed by simply considering gains and losses when going from one state of the molecule to another. In comparing the two expressions, the $^3\Sigma^+$ expression will gain:

$$\epsilon_{6\sigma}^N + \sum_{i=1}^{4} (2J - K)_{6\sigma i\sigma} + (J - K)_{6\sigma 5\sigma} + (4J - 2K)_{6\sigma 1\pi}$$

and $^1\Sigma^+$ will lose

$$\epsilon_{5\sigma}^N + \sum_{i=1}^{4} (2J - K)_{5\sigma i\sigma} + J_{5\sigma 5\sigma} + (4J - 2K)_{5\sigma 1\pi},$$

but

$$\epsilon_{6\sigma}^{\mathrm{SCF}} = \epsilon_{6\sigma}^N + \sum_{i=1}^{5} (2J - K)_{6\sigma i\sigma} + (4J - 2K)_{6\sigma 1\pi}$$

and

$$\epsilon_{5\sigma}^{\mathrm{SCF}} = \epsilon_{5\sigma}^N + \sum_{i=1}^{4} (2J - K)_{5\sigma i\sigma} + J_{5\sigma 5\sigma} + (4J - 2K)_{5\sigma 1\pi}$$

$$\therefore E\,(^3\Sigma^+) - E\,(^1\Sigma^+) = \epsilon_{6\sigma}^{\mathrm{SCF}} - \epsilon_{6\sigma}^{\mathrm{SCF}} - J_{5\sigma 6\sigma}.$$

In this we have again assumed that most electrons are in identical orbitals in both states of the molecule, but reasonable agreement can nevertheless be achieved, as the following table for CO indicates.

CO excitation energies

Excitation	State	Calculated (eV)		Exp. (eV)
	$X\ ^1\Sigma^+$	0		0
$5\sigma \rightarrow 2\pi$	$^3\Pi$	6.17	$a\ ^3\Pi$	6.33
$5\sigma \rightarrow 2\pi$	$^1\Pi$	9.23	$A\ ^1\Pi$	8.51
$1\pi \rightarrow 2\pi$	$^3\Sigma^+$	7.41	$a'\ ^3\Sigma^-$	8.58
$1\pi \rightarrow 2\pi$	$^3\Sigma^-$	9.56	$I\ ^1\Sigma^-$	
$1\pi \rightarrow 2\pi$	$^1\Sigma^-$	9.56	$e\ ^3\Sigma^-$	10.02
$1\pi \rightarrow 2\pi$	$^3\Delta$	8.48	$d\ ^3\Delta$	9.56

6.3 The Brillouin theorem

To improve the wavefunction of a particular state of a molecule it is often necessary to allow a certain mixing with the wavefunctions of other configurations of the same symmetry. If we consider two states, we could write

$$\Psi = a\Psi_0 + b\Psi_1,$$

where Ψ_0 and Ψ_1 are determinantal wavefunctions of the same symmetry type and a and b are the mixing coefficients. The best Ψ will be the mixture which gives the lowest energy, so that we have yet another linear variational problem formally just like a Hückel or SCF problem.

We thus have to solve a set of secular equations such as

$$\begin{aligned} a(H_{00} - E) + bH_{10} \qquad &= 0 \\ aH_{01} \qquad + b(H_{11} - E) &= 0, \end{aligned}$$

where H_{00} is $\langle\Psi_0|\mathrm{H}|\Psi_0\rangle$ and H_{10} $\langle\Psi_1|\mathrm{H}|\Psi_0\rangle$. This looks exactly like a Hückel calculation, but it must be realised that Ψ_0 and Ψ_1 are complete molecular wavefunctions, possibly linear combinations of determinants, and the matrix elements involving them must be taken with the aid of the rules given in Chapter 4.

The solution of the secular equations is performed in the usual manner; the secular determinant is set equal to zero and solved to give the possible values of E which may be substituted back into the secular equations to give the mixing coefficients.

Example. Continuing with the CO molecule, we have a ground state ${}^1\Sigma^+$,

$$\Psi_0 = |(1-5)\sigma^2 1\pi^4| = A;$$

an excited ${}^1\Sigma^+$ state,

$$\begin{aligned} \Psi_1 &= \frac{1}{\sqrt{2}}\{|(1-4)\sigma^2 5\sigma 6\bar{\sigma} 1\pi^4| - |(1-4)\sigma^2 5\bar{\sigma} 6\sigma 1\pi^4|\} \\ &= \frac{1}{\sqrt{2}}(B - C) \end{aligned}$$

The secular determinant will then be

$$\begin{vmatrix} H_{AA} - E & \frac{1}{\sqrt{2}}(H_{AB} - H_{AC}) \\ \frac{1}{\sqrt{2}}(H_{AB} - H_{AC}) & \frac{1}{2}(H_{BB} + H_{CC} - 2H_{BC}) - E \end{vmatrix}$$

The first simplification which can be made to this is to calculate all the diagonal energy terms with respect to the unperturbed energy of the lowest level, i.e. put $H_{AA} = 0$. If the values of all other matrix elements are found using the rules of Chapter 4, it is found that

$$H_{BB} = H_{CC}$$

and

$$H_{AB} = -H_{AC}.$$

Thus the determinant simplifies to

$$\begin{vmatrix} -E & \sqrt{2}H_{AB} \\ \sqrt{2}H_{AB} & (H_{BB} - H_{BC}) - E \end{vmatrix} = 0.$$

From the rule about taking matrix elements between wavefunctions with one spin orbital different, we see that (using the charge-cloud notation)

$$H_{AB} = \epsilon^{N}_{5\sigma 6\sigma} + \sum_{1}^{5} \; [2(5\sigma 6\sigma | i\sigma i\sigma) - (5\sigma i\sigma | 6\sigma i\sigma)] + 4(1\pi 1\pi | 5\sigma 6\sigma) -$$

$$- 2(5\sigma 1\pi | 6\sigma 1\pi).$$

However, using our simplifying trick of using ϵ^{SCF}

$$\epsilon^{\mathrm{SCF}}_{5\sigma 6\sigma} = \langle 5\sigma | \mathrm{H}^{\mathrm{SCF}} | 6\sigma \rangle$$

$$= \epsilon^{N}_{5\sigma 6\sigma} + \sum_{1}^{5} \; [2(5\sigma 6\sigma | i\sigma i\sigma) - (5\sigma i\sigma | 6\sigma i\sigma)] +$$

$$+ 4(1\pi 1\pi | 5\sigma 6\sigma) - 2(1\pi 5\sigma | 1\pi 6\sigma),$$

but the value of non-diagonal matrix elements of the $\mathrm{H}^{\mathrm{SCF}}$ operator is zero, therefore

$$H_{AB} = 0.$$

Thus there is no configuration interaction between the two states. This is a simple example of the Brillouin theorem. For closed-shell states of molecules, mono-excited states do not interact directly with the ground state. However, since mono-excited states can interact with di-excited states, there can be an indirect effect.

Di-excited states may interact with the ground state, as may singly-excited levels of open-shell molecules. For these the technique of configuration interaction followed above is used, but the results may not be so simple.

In general, if a wavefunction is expressed

$$\Psi = aA + bB + cC + dD + \ldots$$

the configuration interaction determinant to be solved will be

$$\begin{vmatrix} -E & H_{AB} & H_{AC} & H_{AD} & H_{AD} \ldots \\ H_{BA} & (H_{BB} - H_{AA} - E) & H_{BC} & H_{BD} \ldots & \\ H_{CA} & H_{CB} & (H_{CC} - H_{AA} - E) & H_{CD} \,. & \\ \cdot & \cdot & \cdot & \cdot & \\ \cdot & \cdot & \cdot & \cdot & \end{vmatrix} = 0.$$

The diagonal element components such as $(H_{BB} - H_{AA})$ can be found exactly as excitation energies, to which they are formally identical, and the off-diagonal elements by using Slater's rules for evaluating matrix elements.

A more general consideration of configuration interaction (often referred to as CI) will be given in chapter 10.

7
OPEN-SHELL SCF METHODS

The majority of molecules of interest in organic or inorganic chemistry have closed-shell ground states. Excited states of these molecules, however, and the ground states of a number of interesting reactive species with an odd number of electrons, tend to have open-shell structures so that open-shell methods are very important.

In Chapter 1 we mentioned the fact that eigenfunctions of an operator which commutes with the hamiltonian lead only to non-zero matrix elements between functions corresponding to the same eigenvalues of this operator. It is usual to insist that electronic wavefunctions are pure spin states since the operator $\mathbf{S}^2$ commutes with the hamiltonian. This is particularly important when calculating expectation values of spin-dependent properties, such as spin-orbit coupling matrix elements. One consequence is that the wavefunction for an open-shell configuration may need to be a linear combination of determinants as explained in the next section.

7.1 Wavefunctions for open shells

For closed shell configurations, the wavefunction will be a single determinant.

$$|\phi_1^2\phi_2^2 \ldots \phi_n^2|,$$

where the spatial parts of the individual ϕ_is form bases for the representation of the symmetry group of the molecule. In the case of heteronuclear diatomic molecules this is the group $C_{\infty v}$ whereas, for example, for H_2CO this is C_{2v}.

Given the character table for the molecule it is a relatively simple matter to write down the orbitals into which electrons are fed. This is clearly explained in many texts on quantum mechanics or group theory.

In particular, for CO there will just be σ- and π-orbitals in the ground state with δ-orbitals etc. in some excited states. The closed shell ground state ($X^1\Sigma^+$) has no resultant spin or orbital angular momentum:

$$|1\sigma^2 2\sigma^2 3\sigma^2 4\sigma^2 5\sigma^2 1\pi^4|.$$

Similarly, we may write the configuration of the ground state of H_2CO as $|1a_1^2 2a_1^2 3a_1^2 4a_1^2 1b_2^2 5a_1^2 1b_1^2 2b_2^2|$ – more will be said about configurations of polyatomic molecules in the next chapter. Returning to CO, consider promoting one of the 5σ-electrons into the 6σ-orbital such that we still have a $^1\Sigma$ state. One of the unpaired electrons will have α-spin and the other will have β-spin so that we have two alternative determinants.

$$|1\sigma^2 2\sigma^2 3\sigma^2 4\sigma^2 5\sigma 6\bar{\sigma} 1\pi^4|$$

and

$$|1\sigma^2 2\sigma^2 3\sigma^2 4\sigma^2 5\bar{\sigma} 6\sigma 1\pi^4|.$$

Neither of these is an eigenfunction of $\mathbf{S}^2$ or S_z – the actual excited singlet wavefunction is the normalized linear combination

$$\frac{1}{\sqrt{2}}\{|(1-4)\sigma^2 5\sigma 6\bar{\sigma} 1\pi^4| - |(1-4)\sigma^2 5\bar{\sigma} 6\sigma 1\pi^4|\}.$$

This result can be obtained using raising and lowering (i.e. step up and step down) operators S^+ and S^-, but the procedure is difficult for the non-specialist. Many programs automatically generate the required linear combinations. However, if we do wish to construct spin states from a given configuration (for example, given orbitals a b c, we may wish to produce a doublet) then there are several ways of projecting out the desired spin functions starting from a single determinant. One such method is described in an appendix.

7.2 Space restrictions

The average interaction for α-spin electrons and β-spin electrons may be different in open-shell systems so that it is not unreasonable that orbitals differing only in their spin quantum number should have different spatial functions. This is the origin of the unrestricted Hartree-Fock (UHF) method implemented in some programs. Unfortunately, UHF wavefunctions are not always eigenfunctions of $\mathbf{S}^2$. If for example we carried out calculations on a molecule with a doublet state and a quartet state that were close in energy, then the equilibrium geometry of the doublet state would be partly characteristic of that for the quartet state.

Although there are methods for projecting out spin eigenfunctions at the end of the UHF calculation, many theoreticians prefer the restricted Hartree–Fock (RHF) method and this is implemented in many open-shell SCF programs. In the RHF method, orbitals differing only in spin quantum number (e.g. $1\pi^+$ and $1\bar{\pi}^+$) have identical spatial parts. In the remainder of this chapter, we shall be concerned only with RHF techniques.

Many of the RHF procedures in common use are based on work by Roothaan; the reader is left to look at the original papers for details of the mathematics for these techniques. The present aim is to present sufficient information for a chemist to use programs that employ Roothaan's open-shell method (or a variant thereof). We shall also mention Nesbet's method.

7.3 Roothaan's open-shell method

We may write the Hartree–Fock equations in matrix form as

$$\mathbf{H}^{\mathrm{SCF}}\mathbf{c} = \mathbf{c}\epsilon^{\mathrm{SCF}},$$

where $\mathbf{c}$ contains the expansion coefficients and ϵ^{SCF} is a matrix of lagrangian multipliers (see Chapter 3); we have assumed orthogonal orbitals so that the overlap matrix $\mathbf{S}$ is equal to an identity or unit matrix. The total wavefunction is invariant to any unitary transformation on the orbitals of a particular symmetry (provided all of these orbitals are equally occupied and all members of a degenerate set occur), so that for a closed-shell molecule we can choose a unitary transformation which makes ϵ^{SCF} a diagonal matrix. The resulting equations now have the simpler form

$$\mathsf{H}^{\mathrm{SCF}}\phi_i = \epsilon_i^{\mathrm{SCF}}\phi_i.$$

As we have seen, this is a pseudo-eigenvalue problem in that the definition of $\mathsf{H}^{\mathrm{SCF}}$ depends on the solutions ϕ_i so that iterative solution is necessary.

For open-shell configurations, the general matrix equation above cannot be simplified in this way and the off-diagonal multipliers remain. However, for a large class of open-shell systems, Roothaan has derived modified SCF hamiltonians, or Fock operators, for closed and open shells so that off-diagonal multipliers between closed (c) and open (o) shells are removed.

The Roothaan–Hartree–Fock method can be used to make stationary (and preferably minimize) an energy expression of the form

$$E = \underbrace{2\sum_k H_k^N + \sum_{kl}(2J-K)_{kl}}_{\text{closed-shell part}} + f\Big[\underbrace{2\sum_m H_m^N + f\sum_{mn}(2aJ-bK)_{mn}}_{\text{open-shell part}} + \underbrace{2\sum_{km}(2J-K)_{km}}_{\substack{\text{open-closed}\\ \text{interaction}}}\Big]$$

Here indices k and l refer to the closed-shell electrons and m and n to the open-shell electrons. a, b, and f are numerical constants which will depend on the particular open-shell problem being considered, f being the fractional occupation of the open shell and a and b differing for different states of the same configuration.

The variation principle can be applied to this expression together with orthogonality constraints and SCF equations of the form

$$\mathsf{F}_c\phi_c = \eta_c\phi_c$$

$$\mathsf{F}_o\phi_o = \eta_o\phi_o$$

result.

It should be noted that the ordering of the η in open-shell calculations is not necessarily the same as the ordering of the one-electron orbital energies ϵ^{SCF}.

The total energy is then equated to

$$E = \sum_k(\epsilon_k^N + \eta_k) + f\sum_m(\epsilon_m^N + \eta_m).$$

The operators F_c and F_o contain not only one-electron operators and coulomb and exchange operators as in the closed-shell case, but in addition constants *a, b,* and *f,* or expressions involving them. These depend on the open-shell case being considered and must normally be specified when writing the data for an SCF program.

There is not complete uniformity about the way in which these extra pieces of data *a, b,* and *f* or their equivalents are used. To a large extent this depends on the authors of the particular program. However, in the input instructions for the available programs there are tables which enable the user to specify the correct combination of constants for the particular case.

The Roothaan method can give an exact solution for many open-shell problems and has been widely used. The method is not applicable to all open-shell configurations; some programs provide for more complicated energy expressions in terms of more general RHF procedures. In Chapter 3, whilst discussing 'convergence control', we mentioned the quantities such as damp factors and level shifters which are used to ensure convergence to a stationary point. For more details, the reader is referred to a paper by Guest and Saunders or to the relevant computer manuals. Convergence problems are more likely in open-shell cases than in closed-shell calculations.

A term used in some programs is 'symmetry equivalence'. If we consider a calculation on a diatomic molecule with configuration $1\pi^1$ then there will be a difference if we use real basis functions and orbitals (i.e. π_x, π_y) rather than complex quantities (i.e. π^+, π^-) since the configuration is neither $1\pi_x$ nor $1\pi_y$ but a linear combination of these. If real basis functions are used in such a case, then the energy expression to be minimized is slightly more complicated but can be handled provided there is suitable 'symmetry equivalencing'.

7.4 Nesbet's method

A rather simple and convenient method of dealing with open-shell electronic configurations which has found favour with organic photochemists is that due to Nesbet. The method involves the use of symmetry and equivalence restrictions. That is to say all orbitals of the same sub-shell, a particular $\phi_{i\lambda}$, have the same radial part in their wavefunction. The self-constant field hamiltonian, $\mathsf{H}^{\mathrm{SCF}}$, which is used is an arbitrary effective hamiltonian for both closed- and open-shell electrons of a given symmetry type. We compute a single set of orbitals for both open and closed shells of the same symmetry with off-diagonal ϵ_{ij} being automatically zero. The total wavefunction is built up from orthonormal doubly and singly occupied orbitals which satisfy the symmetry requirements.

An effective hamiltonian is written using what may be quite arbitrary averaging procedure to give a set of SCF equations. The energy will not normally come out to be expressable as a sum of ϵ^N and ϵ^{SCF} terms but should be close to this. This method is perhaps best understood by considering some simple examples.

As a first example let us consider HeH, with the configuration $1\sigma^2 2\sigma$. We may write the hamiltonian quite arbitrarily as

$$\{\mathsf{H}^N + 2\mathsf{J}_{1\sigma} - \mathsf{K}_{1\sigma} + \mathsf{J}_{2\sigma} - \mathsf{K}_{2\sigma}\}.$$

Hence

$$\epsilon_{1\sigma}^{\mathrm{SCF}} = \epsilon_{1\sigma}^N + J_{1\sigma 1\sigma} + J_{1\sigma 2\sigma} - K_{1\sigma 2\sigma}$$

and

$$\epsilon_{2\sigma}^{\mathrm{SCF}} = \epsilon_{2\sigma}^N + 2J_{1\sigma 1\sigma} - K_{1\sigma 2\sigma}.$$

The energy

$$E = 2\epsilon_{1\sigma}^N + \epsilon_{2\sigma}^N + J_{1\sigma 1\sigma} + 2J_{1\sigma 2\sigma} - K_{1\sigma 2\sigma},$$

but

$$(\epsilon_{1\sigma}^{\mathrm{SCF}} + \epsilon_{1\sigma}^N) + \tfrac{1}{2}(\epsilon_{2\sigma}^{\mathrm{SCF}} + \epsilon_{2\sigma}^N) =$$

$$2\epsilon_{1\sigma}^N + J_{1\sigma 1\sigma} + J_{1\sigma 2\sigma} - K_{1\sigma 2\sigma} + \epsilon_{2\sigma}^N + J_{1\sigma 2\sigma} - \tfrac{1}{2}K_{1\sigma 2\sigma} = E + \tfrac{1}{2}K_{1\sigma 2\sigma}.$$

Thus to obtain the electronic energy we have to add the orbital and one-electron energies but also subtract the molecular integral $\frac{1}{2}K_{1\sigma 2\sigma}$. Now let us consider, as a quite complicated example, an excited state of CO with the wavefunction

$$|1\sigma^2 2\sigma^2 3\sigma^2 4\sigma^2 5\sigma^2 1\pi^+ 1\bar{\pi}^+ 1\pi^- 6\sigma|,$$

a single-determinantal $^3\Pi$ level.

The general form of the SCF equations given previously is

$$\left\{\mathsf{H}^N + \sum_j \mathsf{J}_j - \sum_j{}' \mathsf{K}_j\right\}\phi_i(1) = \epsilon_i^{\mathrm{SCF}}\phi_i(1).$$

For convenience when taking matrix elements the two-electron part of this may again be more specifically written out in a block, but now there is some arbitrariness about this. The Js and Ks in the block are the coulomb and exchange operators.

Occupied orbitals	$1\text{–}5\sigma^2$	6σ	$1\pi^+1\bar{\pi}^+1\pi^-$
σ	$2J-K$	$J-K$	$3J-2K$
π^+	$2J-K$	$J-K$	$3J^0-K^2-K^0$
$\bar{\pi}^+$	$2J-K$	J	$3J^0 \quad -K^0$
π^-	$2J-K$	$J-K$	$3J^0-K^2-K^0$
Average π-electron	$2J-K$	$J-\frac{2}{3}K$	$3J^0-\frac{2}{3}K^2-K^0$

Other average hamiltonians could be written, only each would give a slightly different energy expression. If we set up the SCF equation as just written, how would the energy expression turn out? To answer this we first have to find the energy of the state using the electronic hamiltonian, i.e. use the rules for taking the matrix element of the wavefunction with itself.

$$E = \langle\Psi|\mathrm{H}|\Psi\rangle$$

$$= \sum_{i=1}^{5} 2\epsilon^N_{i\sigma} + 3\epsilon^N_{1\pi} + \epsilon^N_{6\sigma} + \sum_{i,j=1}^{5} (2J-K)_{i\sigma j\sigma} +$$

$$+\sum_{i=1}^{5}(6J-3K)_{i\sigma 1\pi} + \sum_{i=1}^{5}(2J-K)_{6\sigma i\sigma} + 3J^0_{1\pi 1\pi} - K^2_{1\pi 1\pi} +$$

$$+ 3J_{6\sigma 1\pi} - 2K_{6\sigma 1\pi}.$$

We could then use the average SCF hamiltonian we have just written above to express the ϵ^{SCF} in terms of integrals,

$$\text{i.e.}\quad \epsilon^{\mathrm{SCF}}_{i\sigma} = \epsilon^N_{i\sigma} + \sum_{j=1}^{5}(2J-K)_{i\sigma j\sigma} + J_{6\sigma i\sigma} - K_{6\sigma i\sigma} + (3J-2K)_{i\sigma 1\pi}$$

$$\epsilon^{\mathrm{SCF}}_{6\sigma} = \epsilon^N_{6\sigma} + \sum_{i=1}^{5}(2J-K)_{i\sigma 6\sigma} + 3J_{1\pi 6\sigma} - 2K_{1\pi 6\sigma}$$

$$\epsilon^{\mathrm{SCF}}_{1\pi} = \epsilon^N_{1\pi} + \sum_{i=1}^{5}(2J-K)_{1\pi i\sigma} + J_{1\pi 6\sigma} - \tfrac{2}{3}K_{1\pi 6\sigma} + 2J^0_{1\pi 1\pi} - \tfrac{2}{3}K^2_{1\pi 1\pi}$$

$$\therefore\ E(^3\Pi) = \sum_{i=1}^{5} (\epsilon^N + \epsilon^{\text{SCF}})_{i\sigma} + \tfrac{1}{2}(\epsilon^N + \epsilon^{\text{SCF}})_{6\sigma} + \tfrac{3}{2}(\epsilon^N + \epsilon^{\text{SCF}}) +$$

$$+ \tfrac{1}{2} \sum_{i=1}^{5} K_{6\sigma i\sigma} + \tfrac{1}{2} \sum_{i=1}^{5} K_{1\pi i\sigma}.$$

In this example the electronic energy can be expressed as the sum of orbital and one-electron energies with some ten molecular exchange integrals added. The energy towards which we converge is not even the true Hartree–Fock energy since an effective hamiltonian has been used. However, the corrections to this (in the above case a few exchange integrals K) are small. This means that although the energy is poor when uncorrected, the wavefunction is close to a true Hartree–Fock solution and sufficiently accurate to compute expectation values including the correction terms to the energy.

This method is extremely simple to use and can be used for any open-shell case. There are also snags here however. The averaging procedure is rather arbitrary and it is not always clear what the best choice is, particularly if the molecular wavefunction is not a single determinant. Further, the Brillouin theorem no longer holds, so that mono-excited states may not have zero matrix elements with the calculated state even if such matrix elements are zero using the correct hamiltonian. This means that lengthy configuration–interaction calculations are often necessary. The method has been employed in some important applications to photochemistry by Salem.

7.5 Supermatrices

The use of supermatrices is not restricted to open-shell problems and, indeed, the advantages are greater for closed-shell studies, but this is a convenient point briefly to mention them. Coulomb, exchange, and electron-repulsion supermatrices are essentially transformed forms of the relevant two-electron integrals. In an SCF procedure, the two-electron interaction matrix **G** can either be calculated directly from the list of two-electron integrals or from supermatrices. The latter have definite computational advantages in many cases and can lead to considerable improvement in the time taken for SCF calculations. Supermatrix methods are particularly important in view of the new generation of computers (supercomputers) which are presently capable of speeds approaching 135 million floating point operations per second (M flops) for matrix multiplication. The supermatrices facilitate the computation of **J**, **K** and $\mathbf{G} = 2\mathbf{J} - \mathbf{K}$ matrices from one-particle density matrices.

8
CALCULATIONS ON POLYATOMIC MOLECULES

Some of the available computer programs make no real distinction between linear and non-linear molecules; some can only treat linear systems whereas others have been specifically written for larger polyatomic species. The extent to which molecular symmetry is utilized is one of the major distinguishing features between those programs which are suitable for polyatomics.

In this chapter we shall look at the sort of information likely to be required for a calculation; sometimes this will be from the point of view of a program which makes large-scale use of symmetry. However, whether or not the program utilizes molecular symmetry considerations, it is important to be able to understand the symmetry properties of the molecular orbitals output by the program. A number of examples are discussed in Chapter 9.

Before proceeding further, it is necessary to give a brief account of the configurations of polyatomic molecules.

8.1 Configuration of polyatomic molecules

An electronic state of a polyatomic molecule is classified according to the symmetry properties of its wavefunction. The total electronic wavefunction of a given state must transform according to one of the irreducible representations of the point group of the molecule.

These irreducible representations or symmetry species are normally labelled using the nomenclature introduced by Placzek and Mulliken. The symbols A and B represent one-dimensional irreducible representations, the A-representation being symmetric and the B-representation antisymmetric with respect to rotation about the principal axis of symmetry. Two-dimensional representations are labelled E and three-dimensional representations T or F. A prime (′) is added if the species is symmetric with respect to reflection in a plane of symmetry perpendicular to the principal axis and a double prime (″) if it is antisymmetric. If the molecule possesses a centre of symmetry, the suffixes g and u are used to distinguish between species which are symmetric and antisymmetric with respect to inversion at this centre, as for homonuclear diatomic molecules. If there are other elements of symmetry, then the most symmetric species is given the suffix 1 and the other species are given the suffixes 2, 3, etc., in decreasing order of symmetry. A similar system is used in labelling the molecular orbitals, with lower-case letters instead of upper-case.

In order to determine the symmetry species of the electronic states that may be obtained from a given molecular orbital configuration it is necessary to obtain the direct product of the characters of the molecular orbital species of all the electrons in the molecule. The direct product of the characters is simply the ordered product of the corresponding characters of all the molecular orbital species (here and throughout this chapter we are considering only real representations). The direct product of a given species with itself is always the totally symmetric representation, so closed shells may be disregarded. A completely closed-shell species is automatically the totally symmetric species, necessarily a singlet owing to the Pauli principle.

For open-shell configurations we need only determine the direct product of the partially occupied molecular orbitals. If the configuration contains partially occupied orbitals belonging to two- or three-dimensional symmetry species then the direct product may not be one of the irreducible representations of the point group of the molecule. However, the set of characters of the reducible representation may be decomposed into a linear combination of the characters of the component irreducible representations by inspection of the character table of the group. For example, for a molecule having C_{3v} symmetry and a doubly occupied e-orbital, states of symmetry A_1, A_2, and E can be obtained. If the e-orbital contains three electrons then there is only one state, an E-state, which results. Tables giving the symmetry species of the states which may be obtained from various configurations of molecules belonging to the most typical point groups have been produced by Herzberg.

8.2 Input data

The various commonly available programs differ very much in specific details of input but there exist general similarities. Note that some programs actually consist of a suite of closely linked programs – for example, separate integral evaluation and SCF programs. Frequently, the input data for an *ab initio* molecular orbital package is entered into, and is stored in, some interactive computing system; it is then combined with instructions relating to the loading of the program into core and relating to the utilization of system devices and is then submitted to a batch processing system.

The actual form of the data varies greatly from program to program. Some use fixed format input where numbers (e.g. nuclear charges) must be in specific columns of each line; others use 'NAMELIST' input with keywords, e.g. ZETA = 1.534, NAME = 'BUTADIYNE': still others use 'DIRECTIVES' such as TITLE, GEOMETRY, or START with other data in fairly free format. Each of these forms has advantages and disadvantages and each has their devotees – none is difficult to master.

Having loaded an integral evaluation program (or having informed a general program that we wish to calculate atomic integrals) there are various pieces of information needed. We will need to specify details concerning the basis set. For STOs it is necessary to give the quantum numbers n, l, and m and the zeta values. Except for very small systems use of STOs would involve a huge amount of computational effort so that GTOs are used (see Chapter 5). For cartesian gaussians $x^l y^m z^n e^{-\alpha r^2}$ the information required will be the 'type', e.g. S($l + m + n = 0$), P($l + m + n = 1$), D($l + m + n = 2$), etc., and the exponent α. If contractions are being generated then, obviously, contraction coefficients will be needed – frequently, the program will ensure that the contractions are normalized. Some programs provide libraries of suitable basis functions and suitable contractions of GTOs. Given values for nuclear charges, some packages can provide default basis sets.

The other major piece of data needed is the geometry. This might be entered as coordinates or possibly in terms of bond lengths, inter-bond angles, and dihedral angles. Systems which utilize symmetry may need to know details of molecular symmetry and symmetry centres.

Calculations of integrals can now begin. Except for very small molecules or small basis sets it is impossible to keep all the two-electron integrals in the core of the computer. This necessitates the use of system devices such as disks, drums, and tapes. Tapes differ substantially from the other two since their use must necessarily be sequential. Programs exist which make use of the fact that for integrals stored on disks or drums it may be possible to make random access to blocks of integrals and thus achieve greater efficiency. At the end of the SCF procedure it may be desirable to keep the integrals for later use in related systems or for use in other programs. In particular, some packages provide for the calculation of integrals relating to properties such as the dipole moment – these may be useful in a whole range of programs.

Having calculated all the necessary integrals over atomic basis functions, some programs (i.e. those utilizing symmetry) proceed to the construction of integrals over symmetry orbitals; symmetry orbitals will be discussed in Chapter 9.

An SCF procedure can now be started. If these have not already been given in the integral evaluation stage, then the nuclear charges must now be specified. The program will need to know the number of electrons and, particuarly for systems with open shells, details of electronic configuration. For open-shell systems it is frequently necessary to give values for parameters occurring in the energy expression to be minimized or values for 'coupling coefficients' for the hamiltonian – all of these are generally provided in tabulated form for all the common electron configurations (e.g. e^2 giving ^3A). Further information will be necessary if large-scale use is to be made of molecular symmetry.

As described in Chapter 3, some form of initial guess will be needed although some programs provide suitable defaults. Other possible information includes details of convergence thresholds and details of convergence controls – these vary greatly from program to program and are some reflection of the general degree of sophistication.

8.3 Use of symmetry: general considerations

The use of symmetry in the calculation of the integrals over the GTO or STO basis functions can lead to very considerable savings in computer time and can substantially reduce the number of integrals which need to be stored. This, in turn, leads to savings at the SCF stage since reading large numbers of blocks of integrals from system devices can be costly.

We have already seen in Chapter 1 that an integral will be zero if no part of it transforms as the totally symmetric representation of the relevant symmetry group. Thus, for example, in a linear molecule there will be no non-zero integrals between basis functions of σ- and π-symmetry. The use of basis functions representative of the molecular symmetry can lead to important simplifications.

It is also possible to use symmetry in a rather different way. In this case nothing is assumed about the symmetry properties of the basis functions. Suppose that we choose some point in the molecule such as the centre of charge and then try symmetry operations (such as rotations through 90 ° or 180 °) relative to the cartesian axes passing through this point. It will then be possible to identify those two-electron integrals which are equal in magnitude, and thus to calculate only distinct integrals. Programs which implement such procedures often allow the user to declare a number of points that are to be tried as symmetry centres. In this way it is possible to take full advantage of any local symmetry in a part of the molecule.

Use of a basis set of symmetry adapted orbitals can substantially reduce the amount of computer time required for each SCF iteration. However, some authors of SCF programs believe that *constraining the molecular orbitals* to have the correct symmetry does not lead to sufficient time saving to be worth implementing; moreover, some believe that this leads to some reduction in the flexibility of the variational solution. Indeed, it often seems much more pleasing to obtain results which do have all the desired symmetry properties without actually constraining this to be the case.

Clearly, there is a very strong justification for using symmetry considerations when calculating the integrals – even if this is only the use of local symmetry with respect to simple operations. The case for employing molecular symmetry in an SCF procedure can be argued both ways so that many of the commonly available programs differ in this respect.

8.4 Output

This section will necessarily be rather vague because of differences between programs, but we shall attempt to sketch the *sort of output* to be expected from a program for the calculation of *ab initio* molecular orbitals.

In principle, a vast amount of information could be produced; there are frequently options to print, punch on cards, or store on a system device individual atomic or molecular integrals, Fock matrices, overlap matrices, and so on. The quantities of greatest interest to the chemist tend to be energies and coefficients for the m.o.s and the expectation values of any molecular properties which have been computed.

The energies of interest will include the electronic energy of the final wavefunction, the nuclear repulsion energy, and the sum of these – the total energy. The expectation values of the kinetic energy and one-electron hamiltonain operators are often printed. The orbital energies (eigenvalues) may be of interest (see Chapter 6).

It is usual to output the eigenvectors (the coefficients of the basic functions in the m.o.s). Some readers may never have seen a wavefunction in this form and so we shall give an example. If we consider formaldehyde (methanal, H_2CO) with geometry r(C–H) = 0.1102 nm, r(C–O) = 0.1210 nm, θ(H–C–H) = 121.1 °, using the minimal STO basis set, with zeta values

	H	C	O
1s	1.24	5.7	7.7
2s		1.625	2.275
2p		1.625	2.275

then for the closed shell ground state X^1A, with 16 electrons we obtain:

	1	2	3	4
1	−0.000 227 4	−0.005 003 5	0.024 668 4	−0.238 708 5
2	−0.000 227 4	−0.005 003 5	0.024 668 4	−0.238 708 5
3	−0.000 334 1	0.995 761 0	−0.111 562 4	0.168 287 7
4	−0.996 122 8	0.000 063 6	−0.211 542 9	−0.099 485 8
5	0.005 881 1	0.022 584 8	0.283 523 6	−0.604 796 3
6	−0.018 733 3	−0.004 608 7	0.757 338 9	0.455 552 6
7	−0.000 000 0	−0.000 000 0	−0.000 000 0	−0.000 000 0
8	0.0	0.0	0.0	0.0
9	0.005 272 3	0.000 540 9	0.160 389 8	0.219 543 0
10	−0.000 000 0	0.000 000 0	−0.000 000 0	0.000 000 0
11	0.0	0.0	0.0	0.0
12	0.004 849 8	0.001 284 0	−0.172 539 3	0.187 644 7

	5	6	7	8
1	−0.264 975 7	−0.157 389 6	0.0	−0.349 815 7
2	0.264 975 7	−0.157 389 6	0.0	0.349 815 7
3	−0.000 000 0	−0.020 945 2	0.0	−0.000 000 0
4	−0.000 000 0	0.084 624 7	0.0	−0.000 000 0

	5	6	7	8
5	−0.000 000 0	0.068 947 2	0.0	−0.000 000 0
6	−0.000 000 0	−0.491 075 8	0.0	0.000 000 0
7	−0.553 479 8	0.000 000 0	0.0	−0.198 062 2
8	0.0	0.0	0.646 449 5	0.0
9	0.000 000 0	0.462 009 5	0.0	0.000 000 0
10	−0.444 620 4	−0.000 000 0	0.0	0.868 922 6
11	0.0	0.0	0.634 449 1	0.0
12	0.000 000 0	−0.681 411 4	0.0	−0.000 000 0

The numbers at the top of the columns label the molecular orbitals in order of increasing energy. The integers on the left-hand side label the basis functions in the order: H1(1s), H2(1s), C(1s), O(1s), C(2s), O(2s), C($2p_x$), C($2p_y$), C($2p_z$), O($2p_x$), O($2p_y$), O($2p_z$). Note that the coefficients for basis functions (1) and (2) are always of the same magnitude even though we did not constrain them to behave in this way. The coefficients will frequently be stored at the end of the SCF procedure so that other programs requiring these can be linked up to the SCF package. For example, a suite of programs exists for the calculation of spin-orbit coupling effects and these can be interfaced to various SCF programs.

The coefficients, particularly for a large molecule, are often more useful in a slightly modified form as sums of products – the density matrix. If $c_{i\lambda p}$ is the coefficient of the pth basis function of symmetry λ in the ith molecular orbital then element (p, q) of the (partial) density matrix for molecular orbital i is

$$D_{pq}^{\lambda i} = c_{i\lambda p} c_{i\lambda q} n_i,$$

where n_i is the occupancy of orbital i. This matrix can be used, for example, to calculate charges on different nuclei in the highest-occupied (HOMO) and lowest-unoccupied (LUMO) molecular orbitals of organic molecules so as to investigate reactivity. The density matrix is simply the sum of these partial density matrices over molecular orbitals

$$D_{pq}^{\lambda} = \sum_i D_{pq}^{\lambda i} = \sum_i c_{i\lambda p} c_{i\lambda q} n_i.$$

One use of the density matrix is in a Mulliken population analysis which gives the electron densities associated with particular atoms.

The program may also output expectation values of properties such as electric dipole and quadrupole moments, diamagnetic susceptibility and shielding, and electric field gradient.

9
EXAMPLES OF POLYATOMIC CALCULATIONS

For a program which makes widespread use of molecular symmetry, it is clear that in most cases a certain amount of information about the molecule must be obtained before a calculation can begin. We now consider in more detail the information that is required and the methods by which it is obtained. As mentioned in Chapter 8, it is useful to be able to understand this information and the relevant techniques, even if this is not specifically required by the program being used. The molecules chosen as examples are probably much simpler than those that interest the majority of chemists. They should allow us to illustrate the sort of approach that should be employed rather than to discuss in detail the necessary group-theoretical techniques since a number of excellent texts are available.

9.1 H_2O

We shall take as our first example of a calculation of a polyatomic molecule the very simple case of H_2O. In its equilibrium configuration the water molecule possesses a two-fold axis of symmetry along the bisector of the apex angle and two perpendicular planes of symmetry passing through this axis. Hence the molecule belongs to the point group C_{2v}. In order to carry out the calculation as efficiently as possible we require the symmetry orbitals for H_2O in the group C_{2v} and the electronic configuration of the molecule with the molecular orbitals classified according to their irreducible representation in C_{2v}. With a knowledge of the character table of the group C_{2v} this information may readily be obtained by observing the way in which the atomic orbitals on each of the atomic centres transform under the various operations of the group.

The character table for the group C_{2v} is as follows:

	E	C_2	$\sigma_v(xz)$	$\sigma_v(yz)$
A_1	1	1	1	1
A_2	1	1	−1	−1
B_1	1	−1	1	−1
B_2	1	−1	−1	1

The system of axes is chosen according to the usual convention, with the C_2 axis of symmetry as the z-axis and the plane of the molecule as the yz-plane.

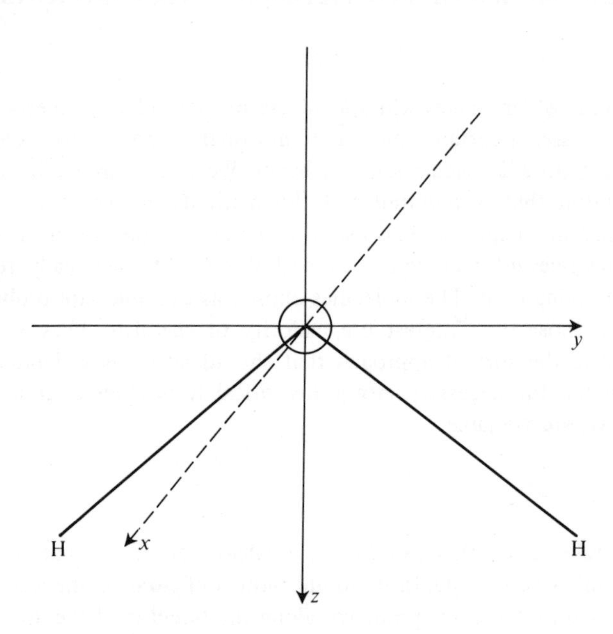

Let us consider first the atomic orbitals centred on the oxygen atom, namely $1s^0$, $2s^0$, $2p_x^0$, $2p_y^0$, and $2p_z^0$. It may be seen at once that the $1s^0$ and $2s^0$ orbitals are not affected by any of the operations of the group C_{2v}. Each of these will therefore be a symmetry orbital belonging to the irreducible representation A_1. Further inspection shows that the $2p_z^0$ orbital is similarly left unchanged by all the operations of the group. The $2p_x^0$ orbital is inverted by the operation of rotation by 180° about the two-fold axis of symmetry (C_2) and by reflection in the plane of the molecule [$\sigma(yz)$]. It therefore belongs to the representation B_1. The $2p_y^0$ orbital is also inverted by the C_2 operation and by reflection in the plane through the symmetry axis perpendicular to the plane of the molecule [$\sigma(xy)$]. This orbital transforms as the irreducible representation B_2. Now we consider the orbitals centred on the hydrogen atoms $1s_1^H$ and $1s_2^H$. It is evident that neither of these orbitals by itself is a symmetry orbital of the group C_{2v} since under the operations C_2 and $\sigma(xy)$, $1s_1^H$ transforms into $1s_2^H$ and vice versa. The correct linear combinations are easily found, however. The (unnormalized) combination $(1s_1^H + 1s_2^H)$ will transform as the irreducible representation A_1 and the antisymmetric combination $(1s_1^H - 1s_2^H)$ will transform as the irreducible representation B_2.

Summarizing, the symmetry orbitals that we have obtained for H_2O are

$$A_1 \begin{cases} 1s^0 \\ 2s^0 \\ 2p_z^0 \\ (1s_1^H + 1s_2^H) \end{cases}$$

$$B_2 \begin{cases} 2p_y^0 \\ (1s_1^H - 1s_2^H) \end{cases}$$

$$B_1 \quad 2p_x^0.$$

The program will therefore build up matrices for each of these representations and diagonalize them separately. From each matrix a set of orbital energies will be obtained. However, in order to carry this out the program will require to know how many orbitals belonging to a given irreducible representation are occupied (and the occupancy of each orbital if there are open shells).

The lowest lying molecular orbital in H_2O will be obviously largely a 1s orbital on the oxygen atom and will therefore have a_1 symmetry. The two symmetry orbitals of B_2 symmetry will give rise to a bonding and antibonding pair of molecular orbitals. Similarly, neglecting 2s–2p hybridization, the 2s and $2p_z^0$ orbitals will separately form bonding and antibonding pairs of molecular orbitals of a_1 symmetry with the symmetric combination $(1s_1^H + 1s_2^H)$. In fact there will be some 2s–2p hybridization, but this will not alter the general picture. The $2p_x^0$ orbital will give a non-bonding orbital of symmetry b_1 as it is the only atomic orbital of this symmetry. H_2O is a ten-electron molecule so, filling the bonding and non-bonding molecular orbitals, there will be three occupied orbitals of a_1 symmetry and one occupied orbital belonging to each of the representations b_2 and b_1. It is useful to have some idea of the composition and relative energy of these orbitals in order to ensure that the SCF iterations converge, although this information is probably not essential for such a simple molecule as H_2O. As stated above, the $1a_1$ orbital will be largely a 1s oxygen atomic orbital and will lie far below the others in energy (~ 20 a.u.). Neglecting 2s–2p hybridization, the $2a_1$ orbital will be largely $2s^0$ and the $3a_1$ largely $2p_z^0$. The $1b_2$ orbital will be mainly $2p_y^0$ and the $1b_1$ orbital will be the non-bonding $2p_x^0$. The $1b_2$ orbital may be predicted to lie slightly below the $3a_1$ orbital in energy as the overlap of the $2p_y^0$ orbital with the hydrogen 1s orbitals would be expected to be somewhat greater than that of $2p_z^0$ at the equilibrium configuration (bond angle 105°). The electronic configuration of H_2O, writing the orbitals in order of increasing energy is therefore

$$(1a_1)^2 (2a_1)^2 (1b_2)^2 (3a_1)^2 (1b_1)^2.$$

The total energy of this configuration expressed as a sum of one-electron terms and coulomb and exchange integrals over molecular orbitals may be obtained using the rules for calculating the matrix elements between Slater determinants given in Chapter 4 and several times illustrated for diatomic molecules.

$$E = \langle |1a_1^2 2a_1^2 1b_2^2 3a_1^2 1b_1^2 | \mathrm{H} | 1a_1^2 2a_1^2 1b_2^2 3a_1^2 1b_1^2 | \rangle$$

$$= \sum_{i=1}^{3} 2\epsilon_{ia_1}^{N} + 2\epsilon_{1b_2}^{N} + 2\epsilon_{1b_1}^{N} + \sum_{i,j=1}^{3} (2J-K)_{ia_1, ja_1} +$$

$$+ 2\sum_{i=1}^{3} (2J-K)_{ia_1, 1b_2} + 2\sum_{i=1}^{3} (2J-K)_{ia_1, 1b_1} + 2(2J-K)_{1b_1, 1b_2}.$$

Using the nomenclature introduced earlier for diatomic molecules, the SCF equations will have the form (λ signifies any one of the symmetry species of the molecule)

$$\left\{ \mathrm{H}_{\lambda}^{N} + \sum_{i=1}^{3} (2\mathrm{J}_{ia_1} - \mathrm{K}_{ia_1}) + (2\mathrm{J}_{1b_1} - \mathrm{K}_{1b_1}) + (2\mathrm{J}_{1b_2} - \mathrm{K}_{1b_2}) \right\} \phi_{\lambda} = \epsilon^{\mathrm{SCF}} \phi_{\lambda}.$$

Expressing the orbital energies as a sum of integrals, as before,

$$\epsilon_{ia_1}^{\mathrm{SCF}} = \left\langle ia_1 \middle| \mathrm{H}_{ia_1}^{N} + \sum_{j=1}^{3} (2\mathrm{J}_{ja_1} - \mathrm{K}_{ja_1}) + (2\mathrm{J}_{1b_2} - \mathrm{K}_{1b_2}) + \right.$$

$$\left. + (2\mathrm{J}_{1b_1} - \mathrm{K}_{1b_1}) \middle| ia_1 \right\rangle$$

$$= \epsilon_{ia_1}^{N} + \sum_{j=1}^{3} (2J-K)_{ia_1, 1b_2} + (2J-K)_{ia_1, 1b_2} + (2J-K)_{ia_1, 1b_1}$$

and

$$\epsilon_{1b_2}^{\mathrm{SCF}} = \left\langle 1b_2 \middle| \mathrm{H}_{1b_2}^{N} + \sum_{i=1}^{3} (2\mathrm{J}_{ia_1} - \mathrm{K}_{ia_1}) + (2\mathrm{J}_{1b_2} - \mathrm{K}_{1b_2}) + \right.$$

$$\left. + (2\mathrm{J}_{1b_1} - \mathrm{K}_{1b_1}) \middle| 1b_2 \right\rangle$$

$$= \epsilon_{1b_2}^{N} + \sum_{i=1}^{3} (2J-K)_{ia_1, 1b_2} + (2J-K)_{1b_2, 1b_2} + (2J-K)_{1b_1, 1b_2}$$

and

$$\epsilon_{1b_1}^{\mathrm{SCF}} = \left\langle 1b_1 \middle| \mathrm{H}_{ib_1}^{N} + \sum_{i=1}^{3} (2\mathrm{J}_{ia_1} - \mathrm{K}_{ia_1}) + (2\mathrm{J}_{1b_2} - \mathrm{K}_{1b_2}) + \right.$$

$$\left. + (2\mathrm{J}_{1b_1} - \mathrm{K}_{1b_1}) \middle| 1b_1 \right\rangle$$

$$= \epsilon^N_{1b_1} + \sum_{i=1}^{3} (2J-K)_{ia_1\,1b_1} + (2J-K)_{1b_1\,1b_2} + (2J-K)_{1b_1\,1b_1}.$$

Therefore, expressing the total energy as a sum of orbital energies, and one-electron energies,

$$E = \sum_{i=1}^{3} (\epsilon^N_{ia_1} + \epsilon^{SCF}_{ia_1}) + (\epsilon^N_{1b_2} + \epsilon^{SCF}_{1b_2}) + (\epsilon^N_{1b_1} + \epsilon^{SCF}_{1b_1}).$$

This again illustrates the general result for closed-shell molecules.

9.2 NH_3

The second example we take is the ammonia molecule in the pyramidal configuration. This example will serve to introduce certain more systematic techniques for obtaining the irreducible representations of the symmetry orbitals and the correct linear combinations of basis orbitals that form the symmetry orbitals. The point group of NH_3 in the pyramidal configuration is C_{3v}. The character table of this group is given below

	E	$2C_3$	$3C_v$
A_1	1	1	1
A_2	1	1	−1
E	2	−1	0

The z-axis is taken to be the C_3 axis of symmetry. The z–x plane includes one of the N–H bonds.

The diagram overleaf represents a view of the three hydrogen atoms, looking down the z-axis.

In this case, the symmetry orbital coefficients cannot be obtained simply by inspection as for H_2O because of the two-dimensional E representation, nor is it obvious which atomic orbitals will form the basis for a given irreducible representation.

To determine the irreducible representation of the symmetry orbitals that can be obtained from a set of symmetry-related atomic orbitals the following procedure is carried out. Apply one of the operations of the group to each of the atomic orbitals of the set in turn. The result of the operation on each orbital will be a linear combination of all the orbitals in the set (some of which will have zero coefficients). Express the set of coefficients obtained in this way in matrix form, thus obtaining the representative of this operation and determine the trace of the matrix (the character of the representative). Repeat this

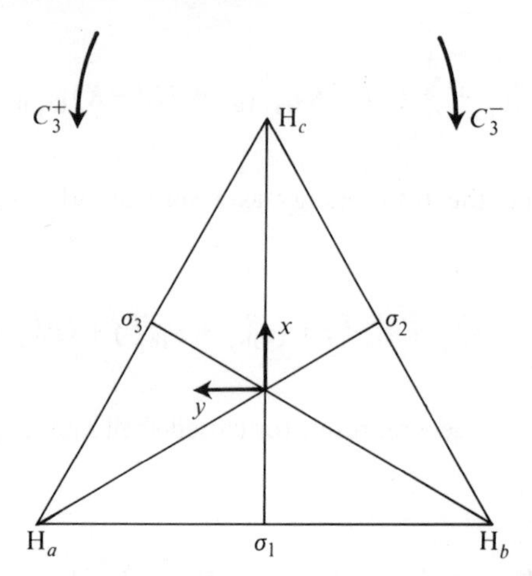

process for all the operations of the group. The resultant set of characters may usually be 'decomposed' by inspection into the characters of its component irreducible representations.

There is usually no need to carry out the whole of this procedure. Very often, on applying an operation, the basis function will be completely unaffected or will simply change sign, thus contributing 1 or −1 to the trace, respectively. The operation may also transform the function entirely into other functions of the set and so for this function the contribution to the trace will be zero. For example, the three symmetry-related 1s hydrogen orbitals are all left unchanged by the identity operation E, and so the character for this operation will be 3. On applying the rotation operation C_3, each of the hydrogen orbitals is transformed into another one of the set, so the character will be zero. Each of the reflection operations σ_v will leave one of the orbitals unchanged and interchange the other two, giving a character of 1. We obtain, therefore, the following set of characters.

E	$2C_3$	$3\sigma_v$
3	0	1

We must now look for the linear combination of the characters of the irreducible representations of the group C_{3v} that will give this particular set of characters. It is readily seen from the character table that the desired combination may be obtained by addition of the characters of the representation E and the representation A_1. These will therefore be the irreducible representations of the symmetry orbitals which may be obtained from the three 1s orbitals. In more complicated cases it may not be possible to obtain the component irreducible representations simply by inspection of the character table of the group. They may be obtained systematically using the formula

$$a_j = \frac{1}{h} \sum_{\mathrm{R}} \chi(\mathrm{R})\chi_j(\mathrm{R}),$$

where a_j is the number of times the jth irreducible representation is contained in the reducible representation (which must be real), $\chi(\mathrm{R})$ is the character of the reducible representation for the operation R, and $\chi_j(\mathrm{R})$ is the character of the jth irreducible representation for the same operation. h is the order (number of operations) of the group.

The derivation of this formula may be found in any of the standard texts on group theory. Instead of decomposing the set of characters obtained for the three 1s hydrogen orbitals by inspection, the irreducible representations could have been obtained by substitution in the above formula. Considering first the A_1 irreducible representation, we obtain

$$a_{\mathrm{A}_1} = \tfrac{1}{6}\{3 \times 1 + 2(1 \times 0) + 3(1 \times 1)\} = 1.$$

Similarly, for the E irreducible representation

$$a_{\mathrm{E}} = \tfrac{1}{6}\{3 \times 2 + 2(0 \times -1) + 3(1 \times 0)\} = 1.$$

Thus the A_1 representation and E representation each occur once in the reducible representation having the set of characters obtained above. There is no need to continue further as these two representations combined give a reducible representation with the correct number of dimensions (3).

The orbitals centred on the nitrogen atom are more easily grouped into their respective irreducible representations. The 1s and 2s atomic orbitals are left unchanged by any operation of the group so they belong to the totally symmetric A_1 representation. Because of our particular choice of z-axis the $2p_z$-orbital is also left invariant under any operation of the group. The $2p_x$- and $2p_y$-orbitals, on the other hand, are transformed by the C_3 rotation into a linear combination of $2p_x$ and $2p_y$. They must therefore form a basis for the two-dimensional E representation.

We must now obtain the coefficients of the linear combinations of atomic orbitals which form the symmetry orbitals of the group. When these coefficients are not obvious by inspection they may be obtained systematically by means of certain operators known as projection operators which project out of any function that part of it belonging to a given irreducible representation. These operators have the form

$$\mathrm{P}(j) = \frac{l_j}{h} \sum_{\mathrm{R}} \chi_j(\mathrm{R})\mathrm{R},$$

where R is one of the operations of the point group of the moelcule. $\chi_j(\mathrm{R})$ is the character of the operation R in the jth irreducible representation. (We again assume that we are considering only real representations).

l_j is the dimension of the jth representation and h is the order (number of operations) of the group. The procedure to obtain the symmetry orbital coefficients of the jth irreducible representation will therefore be as follows: choose one of the basis functions and perform each of the operations of the point group of the molecule on this function in turn. Multiply each of the resulting functions in turn by the character of the jth irreducible representation corresponding to the operation which was used to obtain that function. The sum of the functions obtained in this way is a symmetry orbital of the jth irreducible representation. The numerical factor introduced by the term l_j/h may be omitted as it is replaced by a normalization coefficient by the program. The above procedure will give only one of the components of the set of symmetry orbitals belonging to a degenerate representation. To obtain the other components the process must be repeated for a number of basis functions at least equal to the dimension of the representation. There is a further difficulty for degenerate representations because this method does not in general yield functions that belong to a given row of a degenerate representation, which are the functions that must be supplied to the program. This diffculty can be overcome by deducing the full matrix of the degenerate representation for any one of the operations of the group.

Let us consider once more the 1s orbitals on the three hydrogen atoms of the ammonia molecule. We have seen that from these functions symmetry orbitals belonging to the A_1 and E irreducible representations may be formed. We must now determine the linear combinations of the hydrogen orbitals which belong to these two irreducible representations. Each 1s-orbital is represented by a single basis function labelled $1s_A$, $1s_B$, and $1s_C$ respectively. We use the character table for the group C_{3v}, given above. Applying the operator for the representation A_1 to basis function $1s_A$, we obtain the symmetry orbitals

$$\Psi(A_1) = 1s_A + 1s_B + 1s_C.$$

Applying the projection operator for the E_1 representation to each of the basis functions in turn yields the functions

$$\Psi_1(E) = 2(1s_A) - 1s_B - 1s_C$$

$$\Psi_2(E) = 2(1s_B) - 1s_A - 1s_C$$

$$\Psi_3(E) = 2(1s_C) - 1s_B - 1s_A.$$

These three functions are not linearly independent. We must find two linear combinations of the three functions which will correspond to different rows of the E representation. Let us consider the matrix corresponding to one of three reflection operations (σ_1, σ_2, σ_3). This matrix may be put in diagonal form. As the character of this operation is zero in the E representation the diagonal matrix must be

$$\begin{pmatrix} 1 & 0 \\ 0 & -1 \end{pmatrix}$$

Therefore, one of the required functions will be symmetric with respect to any one of the reflection operations, σ_1, for example, and the other antisymmetric. The symmetric function is just the third of the three functions obtained with the projection operator, namely

$$\Psi_1'(\mathrm{E}) = 2(1s_C) - 1s_B - 1s_A.$$

The antisymmetric function is the difference of the first and second functions

$$\begin{aligned}\Psi_2'(\mathrm{E}) &= (2(1s_A) - 1s_B - 1s_C) - (2(1s_B) - 1s_A - 1s_C)\\ &= 3(1s_A - 1s_B).\end{aligned}$$

The numerical factor may be discarded. Normalization of these functions is usually carried out by the program.

The $2p_x$- and $2p_y$-orbitals on the nitrogen atom also form a basis for the E representation. The σ_1-plane of reflection contains the x-axis. Hence the $2p_x$-orbital will be symmetric with respect to the reflection σ_1 and the $2p_y$-orbital will be antisymmetric. The $2p_x$-orbital therefore belongs with the first of the pair of functions obtained above and the $2p_y$-orbital with the second.

The symmetry orbitals for NH_3 may be summarized as follows:

$$\begin{array}{ll} & 1s^{\mathrm{N}} \\ \mathrm{A}_1 & 2s^{\mathrm{N}} \\ & 2p^{\mathrm{N}} \\ & 1s_A^{\mathrm{H}} + 1s_B^{\mathrm{H}} + 1s_C^{\mathrm{H}} \\ & \left\{\begin{array}{l} 2(1s_C^{\mathrm{H}} - 1s_B^{\mathrm{H}} - 1s_A^{\mathrm{H}} \\ 2p_x^{\mathrm{N}} \end{array}\right. \\ \mathrm{E} & \left\{\begin{array}{l} 1s_A^{\mathrm{H}} - 1s_B^{\mathrm{H}} \\ 2p_y^{\mathrm{N}} \end{array}\right. \end{array}$$

The symmetry orbitals of E symmetry will give one bonding e-orbital. The other occupied molecular orbitals will have a_1 symmetry. Reasonable guesses of the starting coefficients can be made by assuming that the $1a_1$-orbital is almost entirely a nitrogen 1s-orbital and the $2a_1$ and $3a_1$-orbitals are predominantly nitrogen 2s and $2p_z$ respectively, together with a small amount of the hydrogen A_1 symmetry orbital. Assuming that the overlap of the $2p_x$- and $2p_y$-orbitals with the hydrogen orbitals is slightly greater than that of the $2p_z$-orbital, the electronic configuration of NH_3 is:

$$(1a_1)^2\,(2a_1)^2\,(1e)^4\,(3a_1)^2$$

The expression for the total energy of this configuration may be obtained in the same way as for H_2O, using Slater's rules.

9.3 MF_6

As a final example we shall consider the hexafluoride of a transition metal M in an octahedral configuration.

The coordinate system that we shall use is shown in the diagram below.

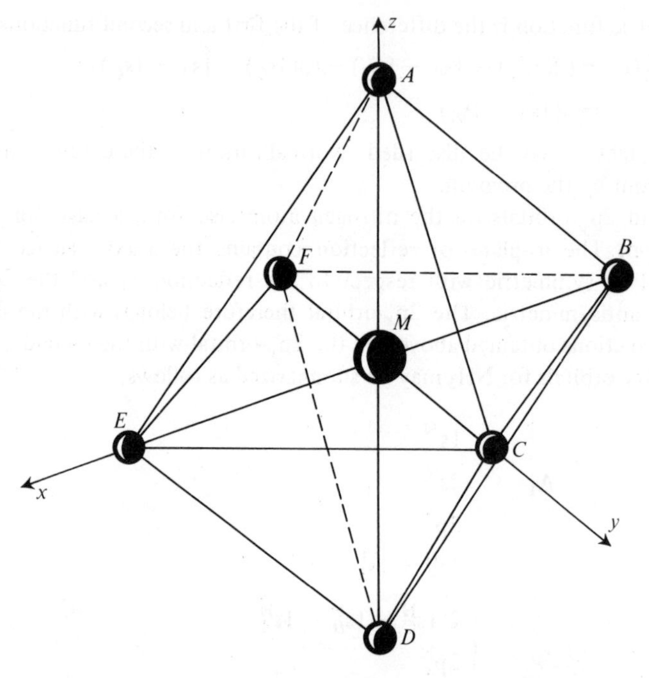

The point group of a molecule in an octahedral configuration is O_h. To determine the symmetry orbitals for this point group we shall need the character table of the group, which is given opposite.

The determination of the symmetry orbitals in this case will be a much longer process than it was for the simple molecules we have just considered, but the same methods can be used.

We consider first the atomic orbitals centred on the transition metal M. The orbitals of s-symmetry are obviously invariant under all the operations of the group and hence belong to the totally symmetric A_{1g} irreducible representa tion. However, it is not obvious to which irreducible representations the p- and d-orbitals belong. Let us take the p-orbitals first. To determine the irreducible representations for which the p-orbitals form a basis we operate on all the p-orbitals with each of the operations of the group in turn and determine the trace of the matrix corresponding to each operation. Instead of treating each type of rotation operation as a separate case we shall consider first the general rotation by an arbitrary angle α.

O_h	E	$8C_3$	$3C_2$	$6C_4$	$6C_2'$	i	$8S_6$	$3\sigma_h$	$6S_4$	$6\sigma_d$
A_{1g}	1	1	1	1	1	1	1	1	1	1
A_{1u}	1	1	1	1	1	−1	−1	−1	−1	−1
A_{2g}	1	1	1	−1	−1	1	1	1	−1	−1
A_{2u}	1	1	1	−1	−1	−1	−1	−1	1	1
E_g	2	−1	2	0	0	2	−1	2	0	0
E_u	2	−1	2	0	0	−2	1	−2	0	0
T_{1g}	3	0	−1	1	−1	3	0	−1	1	−1
T_{1u}	3	0	−1	1	−1	−3	0	1	−1	1
T_{2g}	3	0	−1	−1	1	3	0	−1	−1	1
T_{2u}	3	0	−1	−1	1	−3	0	1	1	−1

The axis of rotation is chosen to be the z-axis, so the p_z- orbital is left unchanged by any rotation about this axis. The p_x- and p_y-orbitals are transformed into linear combinations of p_x and p_y given by the following expressions.

$$C(\alpha)p_x = \cos\alpha\, p_x + \sin\alpha\, p_y$$

$$C(\alpha)p_y = \cos\alpha\, p_x - \sin\alpha\, p_y.$$

The result of the operation $C(\alpha)$ on the p-orbitals may therefore be expressed in matrix form as follows

$$C(\alpha)\; p_x p_y p_z = p_x p_y p_z \begin{pmatrix} \cos\alpha & -\sin\alpha & 0 \\ \sin\alpha & \cos\alpha & 0 \\ 0 & 0 & 0 \end{pmatrix}$$

The character, $\chi[C(\alpha)]$, of the representation for which the p-orbitals form a basis is therefore $(1 + 2\cos\alpha)$.

The operation $S(\alpha)$, a rotation by α followed by a reflection in the xy-plane will invert the p_x-orbital and mix the p_x- and p_y-orbitals in exactly the same way as $C(\alpha)$. Hence the character for this operation will be $(-1 + 2\cos\alpha)$.

The characters for the different rotation operations of O_h may readily be obtained from these general expressions. The characters for the reflection operation (σ) and of inversion at the centre of symmetry (i) may be treated as operations of the type $S(\alpha)$. Thus $\sigma \equiv S(2\pi)$ and $i \equiv S(\pi)$. Hence

$$\chi(\sigma) = -1 + 2 = +1$$

$$\chi(i) = -3.$$

The set of characters obtained for the basis $p_x p_y p_z$ is as follows.

E	$8C_3$	$6C_2$	$6C_4$	$6C_2'$	i	$8S_6$	$3\sigma_h$	$6S_4$	$6\sigma_d$
3	0	−1	1	−1	−3	0	1	−1	1

Inspection of the character table for O_h shows that these are the characters of the T_{1u} irreducible representation. The p-orbitals on M therefore form a basis for this irreducible representation.

We apply the same procedure to determine the irreducible representations for which the d-orbitals of M form a basis. The equations giving the result of a rotation by an angle α of each of the d-orbitals in turn are as follows:

$$C(\alpha)\, d_{x^2-y^2} = \cos 2\alpha d_{x^2-y^2} + \sin 2\alpha d_{xy}$$

$$C(\alpha)\, d_{xy} = -\sin 2\alpha d_{x^2-y^2} + \cos 2\alpha d_{xy}$$

$$C(\alpha)\, d_{xz} = \cos \alpha d_{xz} + \sin \alpha d_{yz}$$

$$C(\alpha)\, d_{yz} = -\sin \alpha d_{xz} + \cos \alpha\, d_{yz}$$

$$C(\alpha)\, d_{z^2} = d_{z^2}.$$

Hence

$$\chi[C(\alpha)] = 2 + 2\cos\alpha + 2\cos 2\alpha$$

and

$$\chi[S(\alpha)] = 2 - 2\cos\alpha + 2\cos 2\alpha.$$

Using these expressions, the set of characters obtained is

E	$8C_3$	$6C_2$	$6C_4$	$6C_2'$	i	$8S_6$	$3\sigma_h$	$6S_4$	$6\sigma_d$
5	−1	1	−1	1	5	−1	1	−1	1

In this case the characters belong to a reducible representation of O_h. By inspection of the character table of O_h it may readily be seen that the above set of characters is a linear combination of the characters for the irreducible representations E_g and T_{2g}.

The orbitals d_{xy}, d_{yz}, d_{xz}, which all have their lobes directed between four fluorine atoms will evidently form a basis for the T_{2g} representation and the d_{x^2} and $d_{x^2-y^2}$ orbitals will form a basis for the E_g representation.

We now consider the symmetry orbitals which can be formed from the s- and p-orbitals of the fluorine ligands. The p-orbitals on the fluorine atoms are chosen so that they are either directed along the MF axis with their positive lobes pointing towards M or perpendicular to this axis. By convention, the p-orbitals directed along the MF axis are labelled σ-orbitals and those perpendicular to the MF axis are labelled π orbitals. It is easily verified that no operation of the group O_h will transform a σ-orbital into a π-orbital and vice versa. The two groups may therefore be treated separately. The s-orbitals on the fluorine atoms belong with the σ-orbitals. It should be pointed out here that most of the computer programs for *ab initio* calculations do not allow any choice in the orientation of the p-orbitals on the ligand atoms. The p-orbitals must in general be oriented so that they point in the same direction as one of the axes of the global coordinate system of the molecule. With this restriction, the p_σ-orbital centred on atom B will be the p_x-orbital centred on B, but the p_σ-orbital on E will be $-p_x$ on E. The p_σ-orbital on F will be the p_y-orbital on F, but the p_σ-orbital on C will be $-p_y$ on C.

The characters of the reducible representation for which the σ-orbitals form a basis may be obtained without difficulty, because in general any operation of the O_h group will either leave a given σ-orbital unchanged or transform it into another member of the set. The orbital will therefore contribute either 1 or 0 to the character of the reducible representation, as explained in the preceding example. The set of characters obtained for the σ-orbitals is

E	$8C_3$	$3C_2$	$6C_4$	$6C_2'$	i	$8S_6$	$3\sigma_h$	$6S_4$	$6\sigma_d$
6	0	2	2	0	0	0	4	0	2

This set of characters may be decomposed into the sum of the characters of the irreducible representations A_{1g}, E_g, and T_{1u}. The σ-orbitals therefore form a basis for these irreducible representations.

The exact form of the symmetry orbitals belonging to each of these irreducible representations may be obtained by applying the projection operators which were described in the preceding example. Applying the operator for the A_1 representation to the orbital σ_A, we obtain

$$\begin{aligned} P_{A_1}\sigma_A = {} & \sigma_A + (2\sigma_A + \sigma_B + \sigma_C + \sigma_E + \sigma_F) + \\ & + (\sigma_A + 2\sigma_D) + (2\sigma_D + \sigma_B + \sigma_E + \sigma_F) + \\ & + (2\sigma_B + 2\sigma_C + 2\sigma_E + 2\sigma_F) + \sigma_D + \\ & + (2\sigma_D + \sigma_B + \sigma_C + \sigma_E + \sigma_F) + (2\sigma_A + \sigma_D) + \\ & + (2\sigma_A + \sigma_B + \sigma_C + \sigma_E + \sigma_F) + \\ & + (2\sigma_B + 2\sigma_C + 2\sigma_E + 2\sigma_F) \\ = {} & 8(\sigma_A + \sigma_B + \sigma_C + \sigma_D + \sigma_E + \sigma_F). \end{aligned}$$

The numerical factor is replaced by a normalizing coefficient by the program. Similarly, applying the operator for T_{1u} to σ_A we obtain the symmetry orbital

$$\Psi_1(T_{1u}) = (\sigma_A - \sigma_B).$$

This is one component of the set of three symmetry orbitals belonging to the representation T_{1u}. The other two components may be obtained by applying the same operator to the orbitals σ_C and σ_E.

$$\Psi_2(T_{1u}) = (\sigma_B - \sigma_E)$$

$$\Psi_3(T_{1u}) = (\sigma_C - \sigma_F).$$

We could have obtained these symmetry orbitals without using the projection operators by forming linear combinations of the σ-orbitals which have the same symmetry as the p-orbitals on the central atom, because these three orbitals also form a basis for the irreducible representation T_{1u}. The combination $(\sigma_A - \sigma_D)$ has the same symmetry as p_z, the combination $(\sigma_C - \sigma_F)$ has the same symmetry as p_y, and the combination $(\sigma_B - \sigma_E)$ has the same symmetry as p_x.

For the E_g symmetry orbitals we have the same problem as we had for the E symmetry orbitals of NH_3. We can project three linearly dependent symmetry orbitals from which we must form two functions which belong to different rows of the degenerate representation. The two functions obtained in this way are

$$\Psi_1(E_g) = (\sigma_B - \sigma_C + \sigma_E - \sigma_F)$$

and

$$\Psi_2(E_g) = (2\sigma_A - \sigma_B - \sigma_C + 2\sigma_D - \sigma_E - \sigma_F).$$

These two symmetry orbitals could equally well have been obtained by forming linear combinations of the σ-orbitals which have the same symmetry as the $d_{x^2-y^2}$ and d_{z^2}-orbitals on the central atom.

The π-orbitals on the fluorine atoms may be further divided into two sets, labelled π and π' respectively, which are shown in the diagram below.

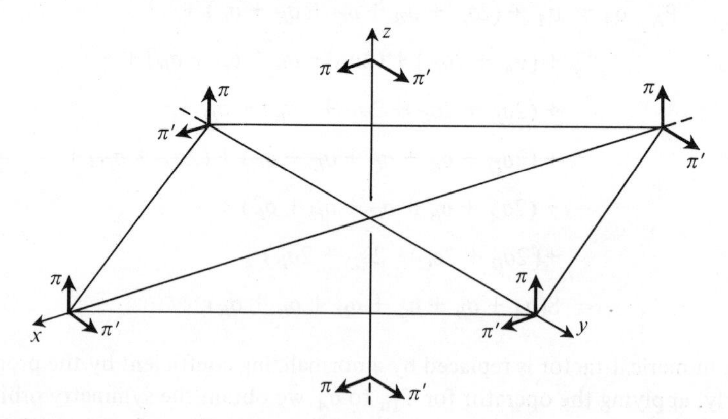

To determine the reducible representation for which the π- and π'-orbitals form a basis we follow the same procedure that we used for the σ-orbitals. This gives the following set of characters:

E	$8C_3$	$3C_2$	$6C_4$	$6C_2'$	i	$8S_6$	$3\sigma_h$	$6S_4$	$6\sigma_d$
12	0	−4	0	0	0	0	0	0	0

This may be decomposed as before to give

$$T_{1g} + T_{2g} + T_{1u} + T_{2u}.$$

With the projection operator for T_{1u} the following functions are obtained.

$$\Psi_1(T_{1u}) = (\pi_C' + \pi_F' + \pi_A + \pi_D)$$
$$\Psi_2(T_{1u}) = (\pi_E' + \pi_B' + \pi_A' + \pi_D')$$
$$\Psi_3(T_{1u}) = (\pi_E + \pi_B + \pi_C + \pi_F)$$

These functions can also be obtained by forming combinations of the π-orbitals which have the same symmetry as the p-orbitals on the central atom.

$$\Psi_1(T_{2g}) = (\pi_E' - \pi_B' + \pi_C' - \pi_F')$$
$$\Psi_2(T_{2g}) = (\pi_E - \pi_B + \pi_A - \pi_D)$$
$$\Psi_2(T_{2g}) = (\pi_C - \pi_F + \pi_A' - \pi_D')$$

The projection operators give for the symmetry orbitals of symmetry T_{1g} and T_{2u}

$$\Psi_1(T_{1g}) = (\pi_C - \pi_F - \pi_A' + \pi_D')$$
$$\Psi_2(T_{1g}) = (\pi_E - \pi_B - \pi_A + \pi_D)$$
$$\Psi_3(T_{1g}) = (\pi_E' - \pi_B' - \pi_C' + \pi_F')$$
$$\Psi_1(T_{2u}) = (\pi_C' + \pi_F' - \pi_A - \pi_D)$$
$$\Psi_2(T_{2u}) = (\pi_E' + \pi_B' - \pi_A' - \pi_D')$$
$$\Psi_3(T_{2u}) = (\pi_E + \pi_B - \pi_C - \pi_F)$$

There are no orbitals of T_{1g} or T_{2u} symmetry on the central atom and so the alternative method is not available to us for these irreducible representations.

In addition to the symmetry orbitals we also need to know the molecular orbital configuration of the molecule in the ground state or in the particular excited state that we wish to calculate. The inner shells of the metal atom and the 1s- and 2s-shells of the fluorine atom may be treated separately, as they take only a negligible part in the bonding. The inner s- and p-shells of the metal atom will have the same order as in the isolated atom and the corresponding m.o. will have a_{1g} or t_{1u} symmetry. The relative position of the orbitals obtained from the 1s- and 2s-orbitals of the fluorine atoms may be determined by comparing the orbital energies of the metal and the fluorine atom obtained by means of Hartree–Fock calculations on the isolated atoms.

The orbitals that take part in the bonding will be the 3d-, 4s-, and 4p-orbitals on the metal atom and the 2p-orbitals on the fluorine atom. The way in which these orbitals interact has been established using qualitative arguments and is usually shown in the form of a diagram similar to that below.

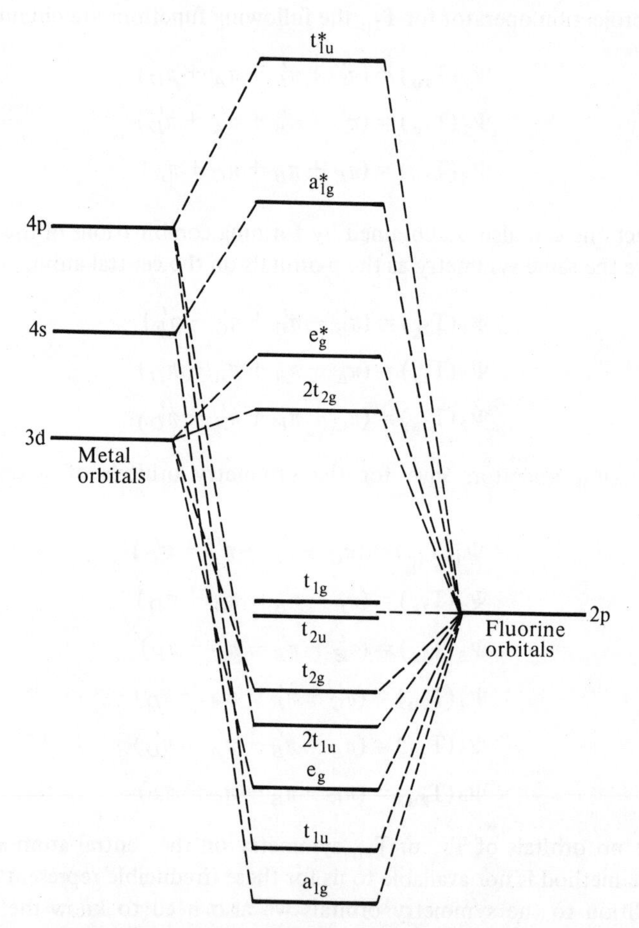

The valence electrons from the metal atom and the fluorine ligands are then allocated to these molecular orbitals, filling each orbital in turn, starting with the orbital of lowest energy. All the orbitals will be completely filled as far as the t_{1g} orbital. The way in which the remaining electrons are distributed between the $2t_{2g}$- and e_g^*-orbitals depends on the difference in energy between these orbitals, so we must use spectroscopic evidence to complete the molecular orbital configuration.

10
ELECTRON CORRELATION

Except for the inclusion of electron spin, no account has been taken so far of relativistic effects in the Hartree–Fock method. Since the virial theorem tells us that inner-shell electrons with the highest potential energy will have the greatest kinetic energy, relativistic corrections may be important for inner electrons in molecules containing heavy atoms. For example, the relativistic energy of a 1s-electron in the magnesium atom is thought to be about 0.2 hartree and that for a 2s-electron about 0.03 hartree.

Relativistic energies are certainly important contributors to the total energy of molecules containing heavy atoms and are difficult to estimate. Frequently we are interested in differences between energy levels rather than absolute energies; the inner electrons are normally unchanged when going from one electronic energy level to another or from one geometry or conformation to another so that the relativistic energy may frequently be ignored.

Rather more serious is the problem of correlation energy and this is the subject of this chapter. It arises from having used a product of one-electron orbitals and having assigned each electron to a particular spatial distribution. In reality, the instantaneous interactions of electrons will not be the same as the average interaction included in the SCF procedure.

10.1 Correlation energy

In Chapter 4 we considered matrix elements of the hamiltonian operator in terms of a sum of one-electron hamiltonians, which we shall denote H_0, and a sum of repulsion terms between pairs of electrons, which we shall denote H':

$$\mathsf{H}^{\mathrm{SCF}} = \mathsf{H}_0 + \mathsf{H}'.$$

For a closed-shell molecule, the exact ground state eigenfunction of H_0 is a normalized antisymmetrized product of one-electron spin-orbitals

$$\Psi_0 = \mathscr{A}\,|\phi_1\phi_2\ldots\phi_n|.$$

Using perturbation theory, the SCF energy to first-order will be

$$\begin{aligned} E^{\mathrm{SCF}} &= \langle\Psi_0|\mathsf{H}^{\mathrm{SCF}}|\Psi_0\rangle \\ &= \langle\Psi_0|\mathsf{H}_0|\Psi_0\rangle + \langle\Psi_0|\mathsf{H}'|\Psi_0\rangle \\ &= E_0 + \langle\Psi_0|\mathsf{H}'|\Psi_0\rangle \end{aligned}$$

The zeroth order energy, E_0, is simply a sum of one-electron orbital energies, $\Sigma\epsilon^N$. We have seen expressions for the first-order energy (i.e. matrix elements of the two-electron part of the hamiltonian in Chapter 4. The second-order (and higher-order) corrections to the energy are collectively known as the correlation energy and arise from the remaining two-body interactions in H'. To be an eigenfunction of $\mathbf{S}^2$, the wavefunction for open-shell systems may be a sum of determinants but the principle is the same.

The correlation energy typically constitutes about one per cent of the total electronic energy so that energies accurate to a few millihartree require the explicit calculation of correlation energy. There are, however, some important applications of *ab initio* wavefunctions for which the Hartree–Fock energies are sufficient.

The following diagram indicates a typical computed potential energy curve of a diatomic molecule compared with the true experimental curve.

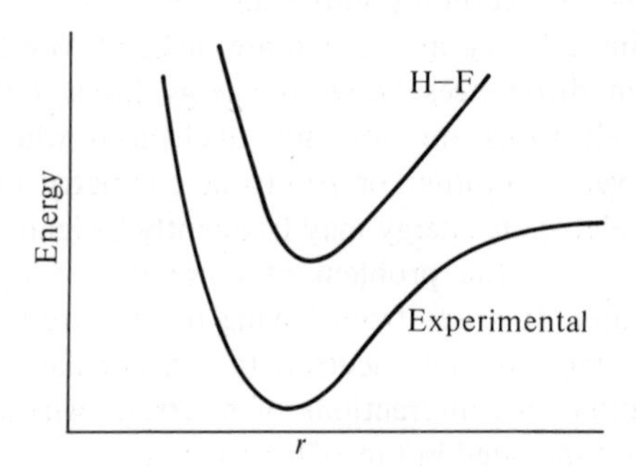

A major fault is the incorrect dissociation behaviour. A Hartree-Fock calculation on H_2 gives H^+ and H^- as products. Electrons tend to stay in pairs in the calculation, making Hartree-Fock calculations particularly inappropriate in computing energy differences where pairing changes as on bond breaking.

Two other things should be noticed. First, the Hartree-Fock curve lies above the experimental, but it is closest to being parallel at the minimum. Secondly, the computed curve is narrower at the bottom of the well; as a result the computed r_e values are normally close to observed results (frequently some hundredths of an ångström unit too small) and the vibration frequencies are too large. The same situation occurs in the multidimensional potential surfaces of polyatomic molecules.

Since computed curves are most realistic at the minimum, the equilibrium geometries are normally rather satisfactorily predicted by polyatomic calculations at the Hartree–Fock level. Furthermore, in cases where one is computing differences in energies, and, where no difference in correlation energy between the two situations is to be expected, *ab initio* computations can give extremely good answers. For instance the repulsive portion of the potential between two rare-gas atoms can be found rather well by calculating the variation of the energy of the He–He system with distance. Since electron correlation arises from residual two-body interactions, the correlation depends largely on pair effects. In the He–He interactions no new pairings of electrons are introduced, so the correlation energy error will be constant with r giving a calculated potential parallelling the true one. Likewise if energies of the ethane molecule are computed at various staggered and eclipsed positions the differences again agree remarkably well with experimental barriers, since there are no differences in pairings of electrons in the various positions (nor are near-degeneracy effects likely to be important).

10.2 Avoided crossings and configuration mixing

The 'avoided crossing rule' tells us that states of diatomic molecules of identical symmetry cannot cross. However, Hartree–Fock SCF curves often do cross each other; this is one of the reasons why we mentioned the mixing of wavefunctions in Chapter 6. If we have two states of the same symmetry with wavefunctions Ψ_1 and Ψ_2 then we can write

$$\Psi = C_1\Psi_1 + C_2\Psi_2,$$

where C_1 and C_2 are coefficients to be found by minimizing the energy. This configuration mixing produces potential curves which do not violate the non-crossing rule.

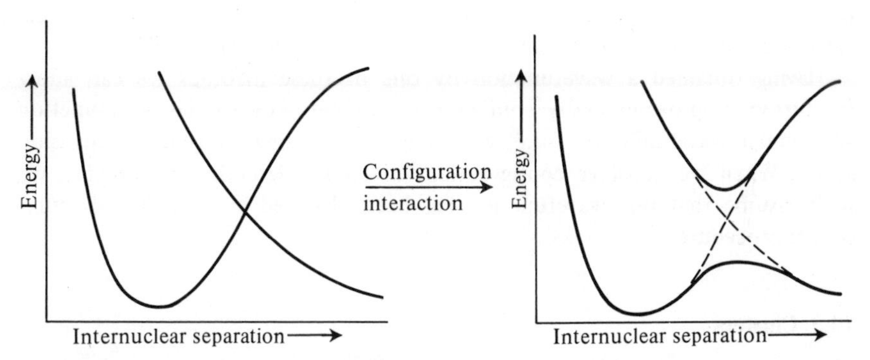

The effects of configuration mixing are most important when states are degenerate, or nearly so.

In general we can write our total wavefunction Ψ as a linear combination of antisymmetrized functions Φ_i built from one-electron spin-orbitals

$$\Psi = \sum_i C_i\Phi_i,$$

where the expansion is as long or as short as we choose.

This short book is about SCF molecular orbital calculations, but this is a convenient point to mention the relationship of various methods to this general expansion.

Valence bond method

The valence-bond (VB) method optimizes the coefficients C_i for a given set of fixed spatial orbitals. Implementations of the VB method tend to be complicated by, for example, matrix elements between non-orthogonal orbitals or difficulties with extensive transformations. VB has had mostly qualitative rather than quantitative success but is now receiving renewed interest. In particular, there have recently been some very convincing results for some light diatomic molecules using a variant known as spin-coupled VB.

Molecular orbital method

As we have seen, the molecular orbital (MO) method truncates the general expansion above at one term and then optimizes the orbitals which make up Φ(or Ψ).

MCSCF

Multiconfiguration SCF procedures use a short expansion in the general expression above. Each Φ_i is built from spatial orbitals which are optimized *and* the coefficients C_i are optimized. Attempts to optimize both the coefficients and the spatial orbitals tend to be rather costly in terms of computer time, even for small molecules such as Li_2, but the results are well worth the effort. By using a suitable set of configurations, it is possible to ensure correct behaviour on dissociation and a smooth transition to the dissociation limit.

Having obtained a wavefunction by one of these methods we can allow for further improvements by configuration mixing between this wavefunction and similar wavefunctions – i.e. we are again extending the general expansion above. When we consider configuration interaction later in this chapter, we shall assume that our wavefunction has been obtained by an SCF moelcular orbital procedure.

10.3 Clusters

We have written the SCF hamiltonian as $\mathsf{H}^{\mathrm{SCF}} = \mathsf{H}_0 + \mathsf{H}'$ where H_0 is a sum of one-electron terms and where H' is a two-electron operator. Interactions between more than two electrons can only occur indirectly. Furthermore, because of the exclusion principle, electrons of the same spin tend to keep apart; this prevents the close approach of more than two electrons. The electron correlation tends to be dominated by two-body electron pair terms.

In this way, the correlation correction to the Hartree–Fock wavefunction can be considered in terms of 'cluster contributions' which simultaneously involve only those electrons in a particular subset of the spin-orbitals – e.g. the four electrons filling a closed π-orbital ($\pi^+ \, \overline{\pi}^+ \, \pi^- \, \overline{\pi}^-$) in a diatomic molecule. These clusters may be classified in terms of the number of electrons and in terms of whether or not there are interactions between clusters. These methods are frequently approached using diagrammatic methods and contributions are classified as linked (connected, irreducible) or unlinked (disconnected, reducible). In other words, an unlinked cluster represents the simultaneous independent effect of two or more smaller clusters.

This cluster approach and diagrammatic perturbation theory methods both lead to the result that the largest linked clusters contribution is from the two-body electron pair terms, as we should expect. Linked cluster three-body terms are much smaller; most higher-order linked clusters can be largely neglected. Two simultaneous independent electron-pair terms constitute a four-body unlinked term – these are the most important of the unlinked clusters.

Pair correlation approaches are likely to be successful considering these comments about clusters. The independent electron pair approximation is not very accurate but methods which allow for interaction between the pairs have been very successful. In particular, the coupled electron pair approximation (CEPA) has produced excellent ground-state properties, such as spectroscopic constants, for diatomic hydrides. Indeed, in some cases, the CEPA results are more reliable than experimental values.

10.4 The configuration interaction method

Configuration interaction (CI) is the most widely used, and probably the most convenient, approach to the correlation-energy problem in large systems. The variationally computed SCF wavefunction is taken, and using its orbitals, several excited states of the appropriate symmetry are constructed. The linear variational method is then used to find the best possible mixing coefficients

$$\Psi = \sum_i C_i \Phi_i,$$

where the first Φ_i term is the SCF wavefunction. This leads to a matrix eigenvalue equation

$$\mathbf{H}\,\mathbf{C} = E\,\mathbf{S}\,\mathbf{C},$$

where **C** contains the coefficients, E is the energy, **H** is the hamiltonian matrix with elements

$$H_{ij} = \langle \Phi_i | \mathsf{H} | \Phi_j \rangle,$$

and **S** is an overlap matrix

$$S_{ij} = \langle \Phi_i | \Phi_j \rangle.$$

It is usual to take the Φ_is, which are called configuration state functions (CSFs), to be orthonormal so that **S** is an identity or unit matrix. It is most convenient to use orthogonal orbitals since we can then calculate matrix elements using

Slater's rules. For non-orthogonal orbitals, rather more complicated expressions, due to Löwdin, would have to be employed.

The CSFs are Slater determinants containing products of spin orbitals. Normally we use spin-adapted CSFs which are eigenfunctions of $\mathbf{S}^2$ and S_z so that for open-shell systems the Φ_is may be a simple sum of determinants. Often they will also satisfy space symmetry conditions – these are symmetry adapted CSFs.

The processes involved in a SCF–CI calculation can be summarized:

(1) Choose basis set and compute integrals. No amount of CI will compensate for a poorly chosen basis set.
(2) Carry out an SCF procedure and obtain orbitals.
(3) Transform the integrals over the basis functions to integrals over molecular orbitals.
(4) Construct suitably adapted CSFs and compute **H** (and **S**).
(5) Find eigenvalues (E) and eigenfunctions (**C**).

We can recognize three extreme categories of correlation between two electrons in a simple bond. First, one electron may remain close to the internuclear axis while the other stays further out. In the second the electrons tend to stay on opposite sides of an axial plane – this is angular correlation. Thirdly, the electrons may tend to stay on different nuclei. For a small CI expansion, we may use chemical reasoning to choose the CSF; note that it is particularly important to incude those CSFs which will allow smooth transition to the correct dissociation limit. Similar considerations hold for MCSCF and VB calculations. This situation will be complicated, of course, by near degeneracy effects and avoided crossings. Choosing CSFs for extensive CI calculations is a more complicated task generally carried out using sophisticated computer programs. CSFs which differ from the SCF configuration by two spin-orbitals tend to be dominant although single excitations are particularly important for open shells (note the Brillouin theorem discussed in Chapter 4). The usual approach is to allow all single and double excitations from the SCF configuration, the 'reference function', within certain restrictions. Methods using more than one reference function are also in common use and are termed 'multiple root' CI.

Inner-shell correlation effects are insensitive to chemical environment so that if we are interested in energy differences we can ignore excitations from inner electrons. Furthermore, excitations to very high virtual (i.e. previously unoccupied) orbitals tend to be less significant than lower energy excitations. Thus it is usual to truncate the virtual space. The CSFs are constructed from both the occupied and unoccupied orbitals generated by the SCF procedure. The latter are also called virtual orbitals because they correspond to a virtual charge in the average field of *all* the electrons. CSFs constructed from virtual orbitals are not a good representation of the part of the wavefunction not represented by the Hartree–Fock limit; this leads to the need for large numbers of CSFs, i.e. slow convergence.

There are many methods for producing suitably adapted CSFs and reducing matrix elements involving these. There has been much recent progress based on representation theories of the permutation or symmetry group and of the general linear or unitary group. The calculation of **H** from integrals can be carried out directly but is often in terms of a 'formula tape' – i.e. CSFs are generated, matrix elements are calculated in symbolic form, and **H** is then computed from these.

The different sizes of cluster are represented in the CI method by different degrees of excitation. Normally only singles and doubles (CISD) are included although for accurate studies, higher excitations can be important. The relative importance of different degrees of excitation has been recently demonstrated for H_2O by allowing *all* excitations of *all* electrons – this is exact solution of the Schrödinger equation (within a particular basis set) and indicates the amount of progress currently being made.

Just as SCF calculations are now being carried out by the non-specialist, programs for the routine calculation of correlation energy should, in time, become available. Indeed, the process has already begun with programs such as GAUSSIAN 80 which can carry out configuration interaction with all double substitutions (CID) from the Hartree-Fock reference determinant using all electrons, or to save computer time, including all but inner-shells. A drawback of the CID wavefunction is that it suffers from an incorrect dependence on the number of particles; it is possible to correct the energy for this so-called 'size-consistency effect'. This program can also include electron correlation at the third-order Møller–Plesset perturbation level. Moreover quadratically convergent MCSCF packages are now being more widely used. It cannot be too long before simple-to-use MCSCF–CI and MCSCF-perturbation-theory packages become freely available for closed- and open-shell molecules of interest to the chemist.

11
APPLICATIONS

We appear to be at the dawn of what may be called 'the third era of quantum chemistry'. During the first era, which predated computers, qualitative agreement between experiment and semi-empirical calculations was sought. The aim was to achieve explanations of the orders of magnitude of experimentally determined quantities. Computers launched a second age with *ab initio* calculations which provided semi-quantitative agreement. Simple explanations were lost but the computations were of real value when experiments were not possible. Now an era of quantitative agreement between calculation and experiment has begun. When we can achieve experimental accuracy with theory then the question of whether it is better to do calculations or experiments is worth debating for any particular problem.

The increasing number of publications, witnessed by the growth of our *Bibliography* of applications of *ab initio* molecular wavefunctions, makes it difficult to summarize all the work in a short chapter, so here we hope to give something of the flavour and range rather than a comprehensive review.

Despire variants, most *ab initio* calculations attempt to solve

$$\mathrm{H}\Psi = E\Psi$$

for a molecule, providing as data the nuclear co-ordinates; atomic numbers of nuclei; the number of electrons in the system; and a suitable basis set of atomic orbitals. The output will be an energy, E, and a wavefunction, Ψ, each broken down into individual orbital contributions. Applications can conveniently be sub-divided into those employing the energy and others starting with the wavefunction.

11.1 Applications based on spectroscopic energies

The obvious, and in many ways most important, application of energy calculations is to spectroscopy which provides a direct measure of the differences in energy between two energy levels. Table 11.1 reminds us of the relationship between the various energy units.

Spectroscopic energy levels or term values (E) are conventionally broken down into a sum of electronic energy (T); vibrational energy (G); rotational energy (F); and fine-structure terms, denoted here as L

$$E = T + G + F + L.$$

Table 11.1
Units and approximate conversion factors

1 atomic unit of energy (hartree)	=	27.21 eV
1 electron volt	=	8065.5 cm^{-1}
1 wave number	=	29 979.25 MHz
1 atomic unit of distance (bohr)	=	0.529 177 Å
	=	$5.291\,77 \times 10^{-11}$ m

There is something paradoxical about the typical calculations of these different types of energy contribution as summarized in the figure.

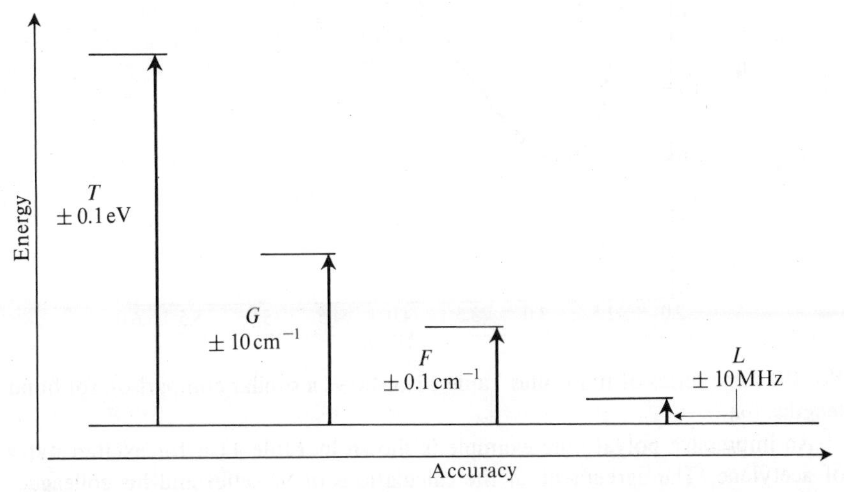

The smaller the energy splitting, the more accurate the calculations appear to be.

Electronic energies used in predictions of excited electronic states are accurate to about ± 0.1 eV for standard good calcultions, although better can be achieved if extreme care is taken over the computation of correlation energies. The figure overleaf gives one real example of such work on SiO in which it may be seen that the calculated curves (solid lines) agree with experimental Rydberg-Klein-Rees potential energy curves (dotted lines) for the ground state but are less good for the excited states.

The ground-state curve agrees well with experiment and even for excited electronic states the curves tend to parallel the truth even if displaced. This is the reason why both vibrational and rotational energy level differences are calculated so much more accurately.

Table 11.2 gives some examples of calculations of vibration frequencies.

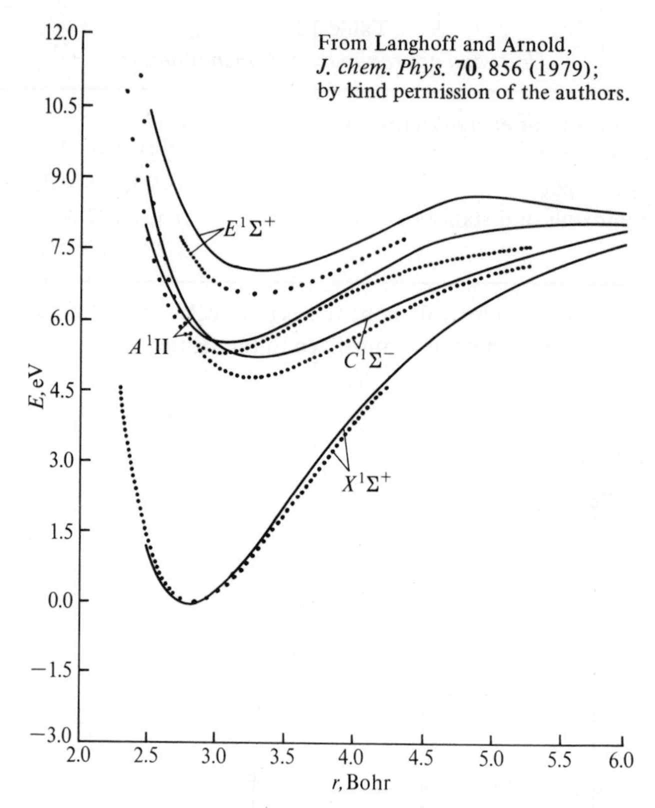

For the same series of molecules Table 11.3 shows a similar comparison for bond lengths.

An impressive polyatomic example is shown in Table 11.4 for excited states of acetylene. The agreement of the calculations of Schaefer and his colleagues with experiment was sufficiently close for the experimental authors to write 'the key to identifying the spectrum was provided by the theoretical calculations'.

Calculations leading experiment, rather than the other way round, is even more the case for the tiny spectroscopic splittings observed as fine and hyperfine structure. These calculations often use the electronic wavefunction in a perturbation treatment.

11.2 Ground electronic state calculations based on energy

By far the majority of published *ab initio* calculations concentrate on molecular ground states and many are both boring and trivial. There seems little point in calculating something which can be measured with ease, especially if it is already known, save for testing the method. None the less there is some point in calculating a molecular ground state geometry if the molecule is unstable or the shape

Table 11.2
Ab initio *claculations of vibration frequencies (ω_e, cm^{-1})*

Molecule	Calculated	Experiment
LiH	1401.5	1405.7
BeH	2064.6	2060.8
BH	2352.1	2366.9
CH	2841.7	2858.5
NH	3269.3	3282.1
OH	3743.6	3739.9
HF	4169.3	4138.7
NaH	1172.3	1172.2
MgH	1492.3	1497.0
A1H	1691.7	1682.6
SiH	2034.7	2041.8
PH	2365.9	2380.0
SH	2676.4	2697.0
HC1	2977.2	2991.1

From Meyer and Rosmus, *J. Chem. Phys.*, **63**, 2386 (1975).

Table 11.3
Ab initio *calculations of bond lengths* (Å)

Molecule	Calculated	Experiment
LiH	1.599	1.595
BeH	1.344	1.343
BH	1.238	1.232
CH	1.122	1.120
NH	1.039	1.037
OH	0.971	0.971
HF	0.917	0.917
NaH	1.891	1.887
MgH	1.728	1.730
A1H	1.645	1.646
SiH	1.526	1.520
PH	1.426	1.422
SH	1.344	1.341
HC1	1.278	1.275

From Meyer and Rosmus, *J. Chem. Phys.* **63**, 2536 (1975).

Table 11.4

Rotational constants as observed for the 0–0 *bands of the* $1\ ^3A_2 - 1\ ^3B_2$ *electronic transition, and as calculated for the theoretical equilibrium geometries of the two states.*

Molecule	State	$A_0 - (B_0 + C_0)/2$ obs. (cm^{-1})	$A_e - (B_e + C_e)/2$ calculated (cm^{-1})	A_e	B_e	C_e
C_2H_2	1^3A_2	12.02 ± .08	12.14	13.22	1.12	1.03
	1^3B_2	11.62 ± .08	11.10	12.22	1.17	1.07
C_2D_2	1^3A_2	6.96 ± .05	6.29	7.13	0.89	0.79
	1^3B_2	6.79 ± .05	5.71	6.59	0.93	0.82

Experimental errors are standard deviations of a least-squares fit.
From Wetmore and Schaefer, *J. Chem. Phys.* **69**, 1648 (1978).

is in dispute. The computational methods which are based on gradient methods can give molecular geometries of experimental quality.

Tautomeric ratios have somewhat more point since often only one possible tautomer is readily observed. However, careful geometry optimization of both forms is essential and the results are only valid for the gas phase since solvation energies cannot be calculated sufficiently accurately even with the best methods from statistical mechanics.

Similar remarks can be made about conformational equilibria. Good gas-phase results are possible if geometries are optimized at all conformations and correlation energy is included. This makes any given instance a major effort on its own but, by a happy cancellation of errors, useful crude conformational data are provided by taking energy differences straight from *ab initio* calculations of energy of different forms with identical geometries but different torsion angles. This type of work has been particularly useful in applications to molecular pharmacology.

Whole reaction surfaces may be calculated as a function of nuclear positions for simple but important fundamental chemical reactions. These surfaces may then serve as the basis for molecular dynamics calculations of reaction rates. Such calculations again need to incorporate correlation energy and represent a major activity and usage of computer time at the present.

Service calculations of the potential surface type also include calculation of intermolecular potentials as a function of relative molecular orientations. These two-body potentials may then serve as the basis of statistical thermodynamic studies of such problems as water structure, solvation, surface tension, or other liquid-state investigations. It is fair to say, however, that the tedious nature of the important provision of potentials as a service to other theoreticians has not proved an attractive type of problem, demanding as it does rather unsocially large amounts of computer time.

11.3 Calculations based directly on the wavefunction

The simplest, and at the same time one of the most useful quantities which can be calculated from a molecular wavefunction, is the electron density, $\Psi^*\Psi$. For simple species such as linear or planar molecules, the electron density may be presented quite straightforwardly as a contour diagram. For less symmetrical molecules there is a problem in presenting the readily calculated values of the electronic density in a manner which can be readily comprehended. Formerly this problem was largely avoided by using the crude Mulliken population analysis as a way of indicating a rough electron distribution. All the charge in a population analysis is assigned to a particular nucleus, and density in bonding regions is arbitrarily alloted to one or other nucleus. Slightly more rigorous is a method whereby the values of $\Psi^*\Psi$ are integrated over a defined region of space to give the total density in a particular part of the molecule.

Pseudo-three dimensional perspective plots of charge density data are now frequently displayed using computer graphics. One widely used method is to subtract the electron density for the various free atoms from that for the molecule and thus obtain the change in electron density on bonding; these are density difference plots. This type of work has been developed very much with applications in organic chemistry and biochemistry in mind. In these areas an alternative way of presenting charge density has proved even more appealing. Instead of producing contour diagrams of charge denisty, molecular electrostatic potential is given instead. In the potential diagrams such as the one over, the contours join positions of identical energy of interaction between the molecular electronic distribution and an isolated proton. There are thus positive and negative regions which are crudely related to chemical reactivity with nucleophilic and electrophilic reagents.

For molecular potentials, just as for electron densities, the advent of computer graphics devices, particularly employing colour to distinguish positive and negative, has been a major stimulus to the employment of the diagrams by experimental chemists.

11.4 Applications of the wavefunction with other operators

It is frequently stated in elementary texts that once one has a good wavefunction then it is possible, using the appropriate operator, Ω, to compute any physical observable, ω:

$$\omega = \frac{\langle\Psi|\Omega|\Psi\rangle}{\langle\Psi|\Psi\rangle}.$$

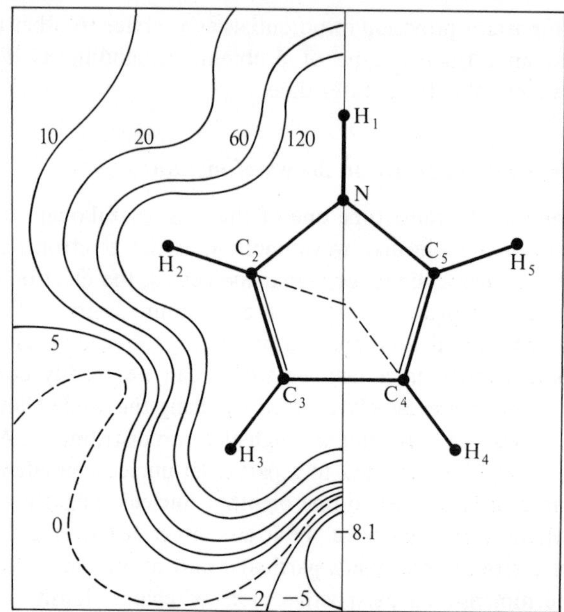

The electrostatic potential field of pyrrole

In practice this is not often done for operators other than the hamiltonian operator which yields energy.

However many of the generally available *ab initio* computer programs will, on demand, provide expectation values of one-electron operators such as $\langle r \rangle$; $\langle 1/r \rangle$; $\langle z \rangle$; $\langle r^2 \rangle$; $\langle x^2+y^2 \rangle$; $\langle z/r^3 \rangle$; or $\langle (3x^2-r^2)/r^5 \rangle$.

These expectation values are related to a number of physical observables which may be calculated to reasonable accuracy. Dipole moments are related to $\langle z \rangle$, quadrupole moments to $\langle z^2-\frac{1}{2}(x^2+y^2) \rangle$, and susceptibilities to $\langle x^2+y^2 \rangle$. The same integrals permit calculation of polarizabilities and field gradients. Unless the wavefunction is of very good quality and includes correlation effects the results are not particularly good.

The same thing may be said for transition probabilities which are related to off-diagonal matrix elements of the dipole operator. With good electronic wavefunctions and vibrational wavefunctions oscillator strengths are in quantitative agreement with experiment.

In some ways more striking success has been achieved with properties which are intrinsically more difficult to compute as a result of being expectation values of two-electron operators. There has been a large number of calculations which use the *ab initio* electronic wavefunctions in a perturbation expression.

In particular off-diagonal matrix elements of the spin-orbit coupling operator are related to spectroscopically observed Λ-doubling and spin-splitting parameters. Computation of these properties has produced spectroscopic splittings correct to within a few megahertz that are of use to radioastronomers trying to assign lines.

Particularly in the case of careful calculations using well-chosen basis-sets and including most of the effects of electron correlation, it is clear that we are entering the third era of *ab initio* computations: calculations of experimental accuracy.

12
CONCLUSIONS

In this short book no attempt has been made at mathematical rigour. Indeed much of the theory, such as group theoretical techniques which are covered in a number of quantum mechanical texts, has been largely ignored, and the reader has been left to go to these texts for such amplification as may be felt necessary. On the other hand, a serious attempt has been made to enable chemists with a normal undergraduate training in elementary wave mechanics to compute *ab initio* wavefunctions for problems that interest them, using one of the published computer programs, and furthermore to utilize the results.

The ready availability of these massive *ab initio* wavefunction programs is due to the generosity of the people who have built and developed them. They provide a striking example of co-operation, which experimentalists might well envy. An experimental worker is frequently loath to allow other scientists, whom he may regard as competitors, to reap benefits of his work in developing apparatus. The writers of programs have not only been ready to furnish copies of their 'apparatus' to interested fellow scientists, but even provide a service of documentation which includes detailed instructions and test input and output.

This organization is the Quantum Chemistry Program Exchange (QCPE) organized from Indiana University (QCPE, Chemistry Department, Room 204, Indiana University, Bloomington, Indiana 47401, USA). Their catalogue now contains over 400 ready-built programs, many of which are highly sophisticated *ab initio* programs capable of treating large polyatomic systems.

The public-spiritedness of these program builders is the reason why a monograph like this is necessary. The big programs are available to people who have the interesting problems to use for themselves. It is too much for the organic chemist to expect theoreticians to do routine computations for him as a service. He should be prepared to run the computations for himself, just as he runs spectra.

The calculation of wavefunctions and energies has been stressed here, since this is the area where most work has been done. Energies of possible configurations of a molecule may be calculated to answer questions about the nature of the ground state of a molecule which may have a profound effect on the thermodynamic properties (by changing the electronic partition function). Similarly the nature of excited states can be clarified by direct *ab initio* calculations.

Other observables can be computed from the wavefunctions by using the appropriate operator. In particular, dipole moments and fine-structure constants have received attention. The computed values are often in very close agreement with the experimentally determined figures, but, more important, they may enable a sign to be given to a quantity that is only ambiguously given by the experiment, and may give an understanding as to the origin of experimental values.

The wavefunction itself is very useful since its square modulus gives a charge density which may be computed at any point in space. The charge densities at nuclei are obviously important in interpreting hyperfine interactions in resonance spectroscopy, but, furthermore, charge densities at a distance from molecules are also interesting. In particular, one can see how the wavefunctions of adjacent molecules in a crystal or polymer might overlap to allow tunnelling of electrons between sites.

Charge densities may be sufficient to explain reaction kinetic behaviour, but *ab initio* calculations can do far more than just provide static indices of charge distributions or electrostatic potentials. Potential surfaces for many reactions have already been calculated. For many of these applications the role of computer graphics in displaying complex output in an intelligible form must be crucial.

In the 1960s and 1970s the development of the use of *ab initio* calculations was directly linked to the advances in large mainframe computers. This is still the source of most published work, but the future is probably linked to developments in two directions, smaller and larger machines.

Minicomputers have brought considerable computing power into a price range where the purchase of a machine is comparable with typical laboratory hardware. Although somewhat slower than machines found in dedicated computer centres, these minicomputers are now being widely used for *ab initio* calculations. By running a machine all night and avoiding the employment of specialist personnel the cost of calculations has fallen considerably. The new generation of microcomputers is certain to increase these possibilities to the extent that the use of *ab initio* methods is likely to grow at an even greater rate in the immediate future.

At the other extreme there is the new generation of parallel and vector processors collectively known as supercomputers. At the present time, these have speeds approaching 135 million floating-point operations per second (megaflops) for matrix multiplication as well as having very fast speeds for scalar operations. With such computers, problems that were previously considered far too expensive are becoming tractable provided they are carefully programmed in such a way as to use the full capability of the hardware. The new generation of computers, coupled with recent theoretical developments, is leading to higher accuracy, particularly in correlation energy studies, and to the possibility of treating much larger systems. There is an interesting review by Guest and Wilson.

There seems to be little doubt that *ab initio* molecular orbital calculations will develop as much in the coming decade as the topic did during the period since the first edition of this book appeared.

APPENDIX

PROJECTED SPIN FUNCTIONS

Producing wavefunctions which are eigenfunctions of spin operators which commute with the hamiltonian becomes a problem for open-shell configurations. Electrons in closed shells or sub-shells constitute no resultant spin, thus the configuration

$$1a_1^2$$

leads to a wavefunction, $|1a_1^2|$, which is a singlet. The excited singlet $1a_1^{\alpha}1b_1^{\beta}$ (alternatively written as $1a_1 1\bar{b}_1$) on the other hand, with one electron having α-spin while the other is of β-spin is only an eigenfunction of $\mathbf{S}^2$ if we take the combination

$$\Psi = \frac{1}{\sqrt{2}}\{|1a_1 1\bar{b}_1| - |1\bar{a}_1 1b_1|\}.$$

The appropriate combinations can be found by use of step-up and step-down operators S^+ and S^-. This procedure is, however, extremely tedious. Fortunately the results are available in the form of tables of spin-functions and particularly useful are the projected spin functions derived by Nesbet.

The details of the mathematical basis of Nesbet's method of producing projected wavefunctions are fully explained in his paper. For the present purpose, which is to present a workable scheme to a non-specialist, it will suffice to give an example of the result and to explain its use.

Let us assume, we have four orbitals outside the closed-shell part of the molecule, indicated by the dashes

$$\underline{1}\ \underline{2}\ \underline{3}\ \underline{4}\ ;$$

each may be a σ- or π^*- or a_1-orbital, etc. If we want the singlet function, two must be of spin α and two of β. We then have the possible determinants

$$\Phi_A = |\alpha\,\beta\,\alpha\,\beta|$$

$$\Phi_B = |\alpha\,\alpha\,\beta\,\beta|$$

$$\Phi_C = |\beta\,\alpha\,\beta\,\alpha|$$

$$\Phi_D = |\beta\,\beta\,\alpha\,\alpha|$$

$$\Phi_E = |\beta\,\alpha\,\alpha\,\beta|$$

$$\Phi_F = |\alpha\,\beta\,\beta\,\alpha|$$

From these we can derive two projected singlet functions.

$$\Psi_{\mathrm{I}} = B + D - E - F$$

$$\Psi_{\mathrm{II}} = A - \tfrac{1}{2}B + C - \tfrac{1}{2}D - \tfrac{1}{2}E - \tfrac{1}{2}F,$$

where B is shorthand notation for Φ_B, etc.

Nesbet's procedure of gaussian elimination produces these functions relatively easily, but also has an additional bonus in reducing the number of terms involved in matrix elements.

If Ψ_{II} were not a projected wavefunction

$$\begin{aligned}\langle\Psi_{\mathrm{II}}|\mathrm{H}|\Psi_{\mathrm{II}}\rangle \text{ would be } H_{AA} &- \tfrac{1}{2}H_{AB} - \tfrac{1}{2}H_{AC} - \tfrac{1}{2}H_{AD} - \tfrac{1}{2}H_{AE} - \tfrac{1}{2}H_{AF}\\ &+ \tfrac{1}{4}H_{BB} - \tfrac{1}{2}H_{BC} + \tfrac{1}{4}H_{BD} + \tfrac{1}{4}H_{BE} + \tfrac{1}{4}H_{BF}\\ &+ H_{CC} - \tfrac{1}{2}H_{CD} - \tfrac{1}{2}H_{CE} - \tfrac{1}{2}H_{CF}\\ &+ \tfrac{1}{4}H_{DD} + \tfrac{1}{4}H_{DE} + \tfrac{1}{4}H_{DF}\\ &+ \tfrac{1}{4}H_{EE} + \tfrac{1}{4}H_{EF}\\ &+ \tfrac{1}{4}H_{FF}\end{aligned}$$

The result of the Nesbet procedure is to produce the function in a conveneint diagram form.

	A	B	C	D	E	F	k_μ
Ψ_{I}	$\frac{1}{2}$	1*	0	1	-1	-1	4
Ψ_{II}	1*	$-\frac{1}{2}$	1	$-\frac{1}{2}$	$-\frac{1}{2}$	$-\frac{1}{2}$	3

Or generally

	A	B	C	D	E	...	k_μ
Ψ_α	$a_{\alpha 1}$	$a_{\alpha 2}$	1*	$x_{\alpha 1}$	$x_{\alpha 2}$	...	$k_{\mu\alpha}$
Ψ_β	$a_{\beta 1}$	1*	$x_{\beta 1}$	$x_{\beta 2}$	$x_{\beta 3}$	...	$k_{\mu\beta}$

The combinations which are functions are read from the starred number to the right, including the starred number, the numbers (x) being the coefficients. The k_μ in the form $k_\mu^{-\frac{1}{2}}$ are normalization constants and numbers (a) to the left of and again including the star are auxiliary coefficients used when taking matrix elements.

If we want the element

$$\langle\Psi_\alpha|\mathrm{H}|\Psi_\beta\rangle$$

this is equal to

$$\left(\frac{k_{\mu\alpha}}{k_{\mu\beta}}\right)^{1/2}\left[\sum_i a_\alpha \sum_j x_\beta \langle\phi_i|\mathrm{H}|\phi_j\rangle\right],$$

index i is read from the left of the diagram up to and including the star and j from and including the star to the right.

Example

$$\begin{aligned}\langle\Psi_{\mathrm{II}}|\mathrm{H}|\Psi_{\mathrm{II}}\rangle &= A\,[A-\tfrac{1}{2}B+C-\tfrac{1}{2}D-\tfrac{1}{2}E-\tfrac{1}{2}F]\\ &= AA-\tfrac{1}{2}AB+AC-\tfrac{1}{2}AD-\tfrac{1}{2}AE-\tfrac{1}{2}AF.\end{aligned}$$

Here

$$AA \equiv \langle\phi_A|\mathrm{H}|\phi_A\rangle, \text{ etc.}$$

Each of these terms AA, AB, etc. must be found using the rules for taking matrix elements between Slater determinants, but clearly there is a great saving of labour.

Similarly,

$$\langle\Psi_{\mathrm{I}}|\mathrm{H}|\Psi_{\mathrm{I}}\rangle = [\tfrac{1}{2}A+B]\ [B+D-E-F]$$

$$\langle\Psi_{\mathrm{I}}|\mathrm{H}|\Psi_{\mathrm{II}}\rangle = \sqrt{(\tfrac{4}{3})}\ [\tfrac{1}{2}A+B]\ [A-\tfrac{1}{2}B+C-\tfrac{1}{2}D-\tfrac{1}{2}E-\tfrac{1}{2}F]$$

or

$$\langle\Psi_{\mathrm{II}}|\mathrm{H}|\Psi_{\mathrm{I}}\rangle = \sqrt{(\tfrac{3}{4})}\ [A]\ [B+D-E-F].$$

As the last two examples show, there is a great simplification in the process of taking matrix elements once one has the table of coefficients x_j and auxiliary coefficients a_i.

The only problem is the production of the table. Nesbet's elegant mathematics has provided a simple means of doing this, and for the benefit of the reader tables for all simple cases are included below.

It should be added that particular care should be taken with Σ states of linear molecules. The Nesbet procedure can only give spin states. When we also need to specify Σ^+ or Σ^- then it may be necessary to add and subtract whole combinations so that the resultant behaves correctly on reflection.

We now consider open-shell wavefunctions of increasing complexity, starting with a single open-shell electron and progressing to the case with five which is as complex as is ever normally likely to be encountered.

1. *One open-shell electron*

We only need to consider the open-shell part of a configuration since the closed-shell part will be totally symmetric.

In this simplest case we only have one possible function = $| \ldots \phi_i^\alpha|$ or the exactly equivalent $| \ldots \phi_i^\beta|$, e.g. $|1\sigma^2 2\sigma|$. This can only give a doublet state wavefunction.

2. *Two open-shell electrons*

(a) Triplet State

The only function will be $| \ldots \phi_j^\alpha \phi_k^\alpha|$', which we will write as

$$A = \alpha\,\alpha$$

$$^3\Psi = A.$$

In every case we only need to consider the wavefunction of one component of a degenerate state, e.g. a triplet and choose here the one with the maximum m_s value since this is the most convenient with which to work.

(b) Singlet State

Here we have no resultant spin and this can have two possible determinants:

$$A = \alpha\,\beta$$

$$B = \beta\,\alpha.$$

Now we can give the simplest example of the Nesbet projected wavefunction

	A	B	k_μ
$^1\Psi$	1*	−1	2

This is interpreted as follows

$$^1\Psi = \frac{1}{\sqrt{2}}\,[\psi_A - \psi_B]$$

$$\langle ^1\Psi|\mathrm{H}|^1\Psi\rangle = H_{AA} - H_{AB}$$

Note that using the projected spin functions we only have two matrix elements instead of four.

3. *Three open-shell electrons*

(a) Quartet State

$$A = \alpha\,\alpha\,\alpha$$

	A	k_μ
$^4\Psi$	1*	1

(b) Doublet States

$$A = \alpha\,\alpha\,\beta$$
$$B = \alpha\,\beta\,\alpha$$
$$C = \beta\,\alpha\,\alpha$$

	A	B	C	k_μ
$^2\Psi_1$	$\frac{1}{2}$	1*	-1	2
$^2\Psi_2$	1*	$-\frac{1}{2}$	$-\frac{1}{2}$	$\frac{3}{2}$

4. *Four open-shell electrons*
(a) Quintet State

$$A = \alpha\,\alpha\,\alpha\,\alpha$$

	A	k_μ
$^4\Psi$	1*	1

(b) Triplet States

$$A = \alpha\,\alpha\,\alpha\,\beta$$
$$B = \alpha\,\alpha\,\beta\,\alpha$$
$$C = \alpha\,\beta\,\alpha\,\alpha$$
$$D = \beta\,\alpha\,\alpha\,\alpha$$

	A	B	C	D	k_μ
$^3\Psi_1$	$\frac{1}{2}$	$\frac{1}{2}$	1*	-1	2
$^3\Psi_2$	$\frac{1}{2}$	1*	$-\frac{1}{2}$	$-\frac{1}{2}$	$\frac{3}{2}$
$^3\Psi_3$	1*	$-\frac{1}{3}$	$-\frac{1}{3}$	$-\frac{1}{3}$	4

(c) Singlet States

$$A = \alpha\beta\alpha\beta$$
$$B = \alpha\alpha\beta\beta$$
$$C = \beta\alpha\beta\alpha$$
$$D = \beta\beta\alpha\alpha$$
$$E = \beta\alpha\alpha\beta$$
$$F = \alpha\beta\beta\alpha$$

	A	B	E	D	E	F	k_μ
$^1\Psi_1$	$\frac{1}{2}$	1*	0	1	−1	−1	4
$^1\Psi_2$	1*	$-\frac{1}{2}$	1	$-\frac{1}{2}$	$-\frac{1}{2}$	$-\frac{1}{2}$	3

5. *Five open-shell electrons*

(a) Sextuplet State

$$A = \alpha\alpha\alpha\alpha\alpha$$

	A	k_μ
$^5\Psi$	1	1

(b) Quartet States

$$A = \alpha\alpha\alpha\alpha\beta$$
$$B = \alpha\alpha\alpha\beta\alpha$$
$$C = \alpha\alpha\beta\alpha\alpha$$
$$D = \alpha\beta\alpha\alpha\alpha$$
$$E = \beta\alpha\alpha\alpha\alpha$$

	A	B	C	D	E	k_μ
$^4\Psi_1$	$\frac{1}{2}$	$\frac{1}{2}$	$\frac{1}{2}$	1*	−1	2
$^4\Psi_2$	$\frac{1}{3}$	$\frac{1}{3}$	1*	$-\frac{1}{2}$	$-\frac{1}{2}$	$\frac{3}{4}$
$^4\Psi_3$	$\frac{1}{4}$	1*	$-\frac{1}{3}$	$-\frac{1}{3}$	$-\frac{1}{3}$	$\frac{4}{3}$
$^4\Psi_3$	1*	$-\frac{1}{4}$	$-\frac{1}{4}$	$-\frac{1}{4}$	$-\frac{1}{4}$	$\frac{5}{4}$

(c) Doublet States

$$A = \alpha\,\beta\,\alpha\,\beta\,\alpha$$
$$B = \alpha\,\beta\,\beta\,\alpha\,\alpha$$
$$C = \alpha\,\alpha\,\beta\,\beta\,\alpha$$
$$D = \beta\,\beta\,\alpha\,\alpha\,\alpha$$
$$E = \beta\,\alpha\,\beta\,\alpha\,\alpha$$
$$F = \beta\,\alpha\,\alpha\,\beta\,\alpha$$
$$G = \beta\,\alpha\,\alpha\,\alpha\,\beta$$
$$H = \alpha\,\beta\,\alpha\,\alpha\,\beta$$
$$I = \alpha\,\alpha\,\beta\,\alpha\,\beta$$
$$J = \alpha\,\alpha\,\alpha\,\beta\,\beta$$

	A	B	C	D	E	F	G	H	I	J	k_μ
${}^2\Psi_1$	$\frac{1}{4}$	$\frac{3}{4}$	$\frac{1}{2}$	$\frac{1}{2}$	1*	1	0	0	1	1	4
${}^2\Psi_1$	$\frac{1}{2}$	$\frac{1}{2}$	0	1*	$-\frac{1}{2}$	$-\frac{1}{2}$	0	-1	$\frac{1}{2}$	$\frac{1}{2}$	3
${}^2\Psi_3$	$\frac{1}{2}$	$\frac{1}{2}$	1*	0	$-\frac{1}{2}$	$-\frac{1}{2}$	1	0	$-\frac{1}{2}$	$-\frac{1}{2}$	3
${}^2\Psi_4$	$\frac{1}{3}$	1*	$-\frac{1}{2}$	$-\frac{1}{2}$	$-\frac{1}{4}$	$\frac{1}{4}$	$\frac{1}{2}$	$-\frac{1}{2}$	$-\frac{1}{4}$	$\frac{1}{4}$	$\frac{9}{4}$
${}^2\Psi_5$	1*	$-\frac{1}{3}$	$-\frac{1}{3}$	$-\frac{1}{3}$	$\frac{1}{3}$	$-\frac{1}{3}$	$\frac{1}{3}$	$-\frac{1}{3}$	$-\frac{1}{3}$	$-\frac{1}{3}$	2

REFERENCES AND FURTHER READING

Altmann, S. L. Group theory, in *Quantum theory*, vol. 2, ed. Bates, D. R., Academic Press, New York, (1962).

Atkins, P. W. *Molecular quantum mechanics*, second edition, Clarendon Press, Oxford (1983).

Bagus, P. S. *Proc. Seminar Selected Topics in Molecular Physics*, IBM, Ludwigsburg (1969).

Bagus, P. S. ALCHEMY studies of small molecules, in *Selected Topics in Molecular Physics*, Verlag Chemie, Weinheim/Bergstrasse (1972).

Binkley, J. S. and Pople, J. A., Møller–Plesset theory for atomic ground state energies. Int. J. Quantum Chem **9**, 229–36 (1975).

Cade, P. E. and Huo, W. M. Electronic structure of diatomic molecules. VI A. Hartree–Fock wavefunctions and energy quantities for the ground states of the first-row hydrides, AH, *J. Chem. Phys.* **47**, 614–48 (1967).

Cade, P. E. and Huo, W. M. Electronic structure of diatomic molecules. VII A. Hartree–Fock wavefunctions and energy quantities for the ground states of the second-row hydrides, AH, *J. Chem. Phys.* **47**, 649–72 (1967).

Cassky, P. and Urban, M. *Ab Initio* calculations, lecture notes in chemistry. Volume 16, Springer-Verlag, Berlin (1980).

Clementi, E. and Roetti, C. Atomic Data and Nuclear Data Tables **14**, 177–478 (1974).

Cook, D. B. Ab initio *valence calculations in chemistry*, Butterworths, London (1974).

Cooper, D. L. and Wilson, S. Universal systematic sequence of even-tempered exponential-type functions in electronic structure studies, in press.

Cotton, F. A. *Chemical applications of group theory*, Wiley, New York (1963).

Daudel, R., Lefèbvre, R., and Moser, C., *Quantum chemistry – methods and applications*, Interscience, New York (1959).

Dunning, T. H. and Hay, P. J. Gaussian basis sets for molecular calculations, in *Methods of electronic structure theory*, ed. Schaefer, H. F., Plenum Press, New York (1977).

Eyring, H., Walter, J., and Kimball, G. E. *Quantum chemistry*, Wiley, New York (1964).

Fletcher, R. and Powell, M. J. D., A rapidly convergent descent method for minimization *Comput. J.* **6**, 163–8 (1963).

Guest, M. F. and Saunders, V. R. On methods for converging open-shell Hartree–Fock wavefunctions, *Mol. Phys.* **28**, 819–28 (1974).

Guest, M. F. and Saunders, V. R. ATMOL3 user manuals, Rutherford Appleton Laboratory, Science and Engineering Research Council, Chilton, Didcot, Oxon, OX11 0QX, UK (1976).

Guest, M. F. and Wilson, S. (editors) Electron correlation, *Proc. Daresbury study weekend* Daresbury Laboratory, Warrington, WA4 4AD, UK (1979).

Guest, M. F. and Wilson, S. The use of vector processors in quantum chemistry; experience in the UK., in *Supercomputers in chemsitry*, ed. Lykos, A. and Shavitt, I., American Chemical Society, Washington, D.C. (1981).

Hinze, J. (editor), *The unitary group*, Lecture notes in chemistry, volume 22 Springer-Verlag, Berlin (1981).

Kutzelnigg, W. Pair correlation theories, in *Methods of electronic structure theory*, ed. Schaefer, H. F., Plenum Press, New York (1977).

Lykos, P. and Shavitt, I. (editors) *Supercomputers in chemistry* American Chemical Society, Washington, D.C. (1981).

McLean, A. D., Weis, A., and Yoshimine, M. Configuration interaction in the hydrogen molecule – the ground state, *Rev. Mod. Phys.* **32**, 211–18 (1960).

McWeeny, R. *Coulson's Valence*, third edition, Clarendon Press, Oxford (1979).

McWeeny, R. and Sutcliffe, B. T. *Methods of molecular quantum mechanics*, Academic Press, London (1969).

Meyer, W. Configuration expansion by means of pseudo-natural orbitals, in *Methods of electronic structure theory*, ed. Schaefer, H. F., Plenum Press, New York (1977).

Meyer, W. and Rosmus, P. PNO – CI and CEPA studies of electron correlation effects. III. Spectroscopic constants and dipole moment functions for the ground states of the first-row and second-row diatomic hydrides, *J. Chem. Phys.* **63**, 2356–75 (1975).

Møller, C. and Plesset, M. S. Note on an approximation for many-electron systems. *Phys Rev.* **46**, 618–22 (1934).

Moskowitz, J. W. and Snyder, L. C. POLYATOM: A general computer program for *ab initio* claculations, in *Methods of electronic structure theory*, ed. Schaefer, H. F., Plenum Press, New York (1977).

Murrell, J. N., Kettle, S. F. A., and Tedder, J. M. *Valence theory*, John Wiley, London (1970).

Murtaugh, B. A. and Sargent, R. W. H. Computational experience with quadratically convergent minimization methods. *Comput. J.* **13**, 185–93 (1980).

Nesbet, R. K. Configuration interaction in orbital theories, *Proc. R. Soc. London, A 230*, 312–21 (1955).

Nesbet, R. K. Construction of symmetry-adapted functions in the many-particle problem, *J. Math. Phys.* **2**, 701–9 (1961).

Pilar, F. L. *Elementary quantum chemistry*, McGraw–Hill, New York (1968).

Pople, J. A. *A priori* geometry predictions in *Applications of electronic structure theory*, ed. Schaefer, H. F., Plenum Press, New York (1977).

Pople, J. A., Binkley, J. S., and Seegar, R. Theoretical models incorporating electron correlation. *Int. J. Quantum Chem. Symposium* **10**, 1–19 (1976).

Pople, J. A., Seeger, R., and Krishnan, R., Variational configuration interaction methods and comparison with perturbation theory *Int. J. Quantum Chem. Symposium* **11**, 149–00 (1977).

Quantum Chemistry Program Exchange *Guide and index to QCPE catalog volume XIII*, 1981. Programs available include GAUSSIAN 70 (and its offspring), POLYATOM, HONDO, MULTIBOND, and many others.

Richards, W. G., Walker, T. E. H., and Hinkley, R. K. *A Bibliography of* ab initio *molecular wavefunctions* Clarendon Press, Oxford (1971).

Richards, W. G., Walker, T. E. H., Farnell, L., and Scott, P. R. *A bibliography of* ab initio *molecular wavefunctions: Supplement for 1970–1973*, Clarendon Press, Oxford (1974).

Richards, W. G. Scott, P. R., Colbourn, E. A., and Marchington, *A. F. A Bibliography of* ab initio *molecular wavefunctions: Supplement for 1974–1977.* Clarendon Press, Oxford (1978).

Richards, W. G., Scott, P. R., Sackwild, V., and Robins, S. A. *A bibliography*

of ab initio *molecular wavefunctions: Supplement for 1978–1980.* Clarendon Press, Oxford (1981).
Richards, W. G., Trivedi, H. P., and Cooper, D. L. *Spin-orbit coupling in molecules.* Clarendon Press, Oxford (1981).
Roos, B. D. and Siegbahn, P. E. M. Gaussian basis sets for the first and second-row atoms *Theoret. chim. Acta* **17**, 209–15 (1970).
Roos, B. O. and Siegbahn, P. E. M. The direct configuration interaction method from molecular integrals, in *Methods of electronic structure theory*, ed. Schaefer, H. F., Plenum Press, New York (1977).
Roothaan, C. C. J. New Developments in molecular orbital theory, *Revs. Modern Phys.* **23**, 69–89 (1951).
Roothaan, C. C. J. Self-consistent field theory for open shells of electronic systems, *Revs. Modern Phys.* **32**, 179–85 (1960).
Saunders, V. R. and Hillier, I. H. A 'level-shifting' method for converging closed shell Hartree–Fock wavefunctions, *Int. J. Quant. Chem.* **7**, 699–705 (1973).
Saxe, P., Schaefer, H. F., and Handy, N. C. Exact solution (within a double-zeta basis set) of the Schrödinger equation for water *Chem. Phys. Lett.* **79**, 202–4 (1981).
Schaefer, H. F. (editor) *Methods of electronic structure theory*, Plenum Press, New York (1977).
Schaefer, H. F. (editor) *Applications of electronic structure theory*, Plenum Press, New York (1977).
Schonland, D. *Molecular symmetry*, Van Nostrand, New York (1965).
Shavitt, I. The method of configuration interaction, in *Methods of electronic structure theory*, ed. Schaefer, H. F., Plenum Press, New York (1977).
Simons, J. P. *Photochemistry and spectroscopy*, Wiley, London (1971).
Slater, J. C. *Electronic structure of molecules*, McGraw–Hill, New York (1963).
Wahl, A. C. and Das, G. The multiconfiguration self-consistent field method, in *Methods of electronic structure theory*, ed. Schaefer, H. F., Plenum Press, New York (1977).

INDEX